Lecture Notes in Artificial Inte

Edited by R. Goebel, J. Siekmann, and W. Wahlster

Subseries of Lecture Notes in Computer Science

Jordi Solé-Casals   Vladimir Zaiats (Eds.)

Advances in Nonlinear Speech Processing

International Conference
on Nonlinear Speech Processing, NOLISP 2009
Vic, Spain, June 25-27, 2009
Revised Selected Papers

 Springer

Volume Editors

Jordi Solé-Casals
Vladimir Zaiats
Department of Digital Technologies and Information
Escola Politècnica Superior, Universitat de Vic
c/. Sagrada Família, 7, 08500 Vic (Barcelona),Spain
E-mail:{jordi.sole, vladimir.zaiats}@uvic.cat

Library of Congress Control Number: 2010920465

CR Subject Classification (1998): I.2.7, I.5.3, I.5.4, G.1.7, G.1.8

LNCS Sublibrary: SL 7 – Artificial Intelligence

ISSN   0302-9743
ISBN-10   3-642-11508-X Springer Berlin Heidelberg New York
ISBN-13   978-3-642-11508-0 Springer Berlin Heidelberg New York

Preface

This volume contains the proceedings of NOLISP 2009, an ISCA Tutorial and Workshop on Non-Linear Speech Processing held at the University of Vic (Catalonia, Spain) during June 25-27, 2009.

NOLISP 2009 was preceded by three editions of this biannual event held 2003 in Le Croisic (France), 2005 in Barcelona, and 2007 in Paris. The main idea of NOLISP workshops is to present and discuss new ideas, techniques and results related to alternative approaches in speech processing that may depart from the mainstream. In order to work at the front-end of the subject area, the following domains of interest have been defined for NOLISP 2009:

1. Non-linear approximation and estimation
2. Non-linear oscillators and predictors
3. Higher-order statistics
4. Independent component analysis
5. Nearest neighbors
6. Neural networks
7. Decision trees
8. Non-parametric models
9. Dynamics for non-linear systems
10. Fractal methods
11. Chaos modeling
12. Non-linear differential equations

The initiative to organize NOLISP 2009 at the University of Vic (UVic) came from the UVic Research Group on Signal Processing and was supported by the Hardware-Software Research Group.

We would like to acknowledge the financial support obtained from the Ministry of Science and Innovation of Spain (MICINN), University of Vic, ISCA, and EURASIP.

All contributions to this volume are original. They were subject to a double-blind refereeing procedure before their acceptance for the workshop and were revised after being presented at NOLISP 2009.

September 2009

Jordi Solé-Casals
Vladimir Zaiats

Organization

NOLISP 2009 was organized by the Department of Digital Technologies, University of Vic, in cooperation with ISCA and EURASIP.

Scientific Committee

Co-chairs:
: Jordi Solé-Casals (University of Vic, Spain)
: Vladimir Zaiats (University of Vic, Spain)

Members:
: Frédéric Bimbot (IRISA, Rennes, France)
: Mohamed Chetouani (UPMC, Paris, France)
: Gérard Chollet (ENST, Paris, France)
: Virgínia Espinosa-Duró (EUPMt, Barcelona, Spain)
: Anna Esposito (Segonda Università degli Studi di Napoli, Italy)
: Marcos Faúndez-Zanuy (EUPMt, Barcelona, Spain)
: Christian Jutten (Grenoble-IT, France)
: Eric Keller (University of Lausanne, Switzerland)
: Gernot Kubin (TU Graz, Austria)
: Stephen Laughlin (University of Edinburgh, UK)
: Enric Monte-Moreno (UPC, Barcelona, Spain)
: Carlos G. Puntonet (University of Granada, Spain)
: Jean Rouat (University of Sherbrooke, Canada)
: Isabel Trancoso (INESC, Lisbon, Portugal)
: Carlos M. Travieso (University of Las Palmas, Spain)

Local Committee

Co-chairs:
: Jordi Solé-Casals (University of Vic, Spain)
: Vladimir Zaiats (University of Vic, Spain)

Members:
: Montserrat Corbera-Subirana (University of Vic, Spain)
: Marcos Faúndez-Zanuy (EUPMt, Barcelona, Spain)
: Pere Martí-Puig (University of Vic, Spain)
: Ramon Reig-Bolaño (University of Vic, Spain)
: Moisès Serra-Serra (University of Vic, Spain)

Referees

Mohamed Chetouani   Pere Martí-Puig   Jordi Solé-Casals
Gérard Cholet   Enric Monte-Moreno   Isabel Trancoso
Virgínia Espinosa-Duró   Ramon Reig-Bolaño   Carlos M. Travieso
Anna Esposito   Carlos G. Puntonet   Vladimir Zaiats
Marcos Faúndez-Zanuy   Jean Rouat

Sponsoring Institutions

Ministerio de Ciencia e Innovación (MICINN), Madrid, Spain
University of Vic, Catalonia, Spain
International Speech Communication Association (ISCA)
European Association for Signal Processing (EURASIP)

Table of Contents

Keynote Talks

Contributed Talks

Multimodal Speech Separation

Bertrand Rivet[1] and Jonathon Chambers[2]

[1] GIPSA-lab, CNRS UMR-5216, Grenoble INP, Grenoble, France
`bertrand.rivet@gipsa-lab.grenoble-inp.fr`
[2] Electronic and Electrical Engineering, Loughborough University, UK
`j.a.chambers@lboro.ac.uk`

Abstract. The work of Bernstein and Benoît has confirmed that it is advantageous to use multiple senses, for example to employ both audio and visual modalities, in speech perception. As a consequence, looking at the speaker's face can be useful to better hear a speech signal in a noisy environment and to extract it from competing sources, as originally identified by Cherry, who posed the so-called "Cocktail Party" problem. To exploit the intrinsic coherence between audition and vision within a machine, the method of blind source separation (BSS) is particularly attractive.

1 Introduction

Have you wondered why it is so difficult to hear the driver of a car from a back seat, especially in a noisy environment? It is well known that the human brain is able to identify and to sort heard sounds: it discriminates the different audio sources and it tries to extract the source(s) of interest. However, when the environment noise is too loud, this faculty becomes insufficient. Thus how can it be explained that in the same noisy conditions you do can understand what your neighbour says ? The answer comes from the cognitive sciences. Your brain merges what you hear and what you see, especially the motion of speech articulatories: speech is multimodal (e.g., [8,33,34]).

Blind source separation (BSS) was introduced in the middle of the 80's and formalised in the begin of the 90's (e.g., [5,13,14,15]). The aim of BSS consists in recovering unknown signals from mixtures of them with as few prior information about sources as possible. BSS can thus be viewed as an extention of speech enhancement.

The aim of audiovisual speech separation is not to mimic the human brain faculties but to propose original multimodal algorithms which exploit the coherence between audio and video speech modalities.

2 Multimodality of Speech

The bimodal nature of speech is now a basic feature, both for the understanding of speech perception [34] and for the development of tools for human–human and human–machine communication [32,21]. It is thus well known that the vision of

J. Solé-Casals and V. Zaiats (Eds.): NOLISP 2009, LNAI 5933, pp. 1–11, 2010.

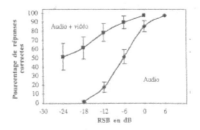

Fig. 1. Recognition rate with audiovisual speech versus audio speech, from [8]

a speaker's face influences what is perceived. One the most famous example is the McGurk's effect [17] which is an audiovisual illusion: superimposing an audio stimulus [ba] with a video stimulus [ga] leads to perceive [da]. Indeed, most of people are able to lip read (e.g., [2,8,33]) even if they are not conscious of doing so: the recognition rate with audiovisual speech increases compared to audio speech only (Fig. 1). In addition, the vision of the speaker's face not only favours understanding but also enhances speech detection in a noisy environment [3,12,16]. All of these studies finally exploit the redundancies and complementarities between visual and audio modalities of speech. Indeed, it seems intuitive that there are strong links between the motion of the speaker's lips and the speech signal. However, this redundancy is not total: there are also complementarities between audio and video modalities. As mentioned in the introduction, the video modality might enhance the perception of the audio modality.

3   Multimodal Speech Separation

Speech enhancement has received much attention due to the many potentials applications: automatic speech recognition, used before a coder designed to clean speech, to improve intelligibility or to extract a specific speaker for instance. Several methods have been proposed in the last decades both in the time domain (e.g., [9]) and in the frequency domain (e.g., [7]). The extension of speech enhancement problem to separation of multiple simultaneous speech/audio sources is a major issue of interest in the speech/audio processing area. This problem is sometimes referred to as the "cocktail party" problem [4] and can be formalised as blind sources separation.

All the facts presented in the previous section leads us to suggest that sound enhancement could exploit the information contained in the coherent visible movements of the speaker's face. The following review presents a quick overview of the proposed audiovisual speech enhancement methods.

3.1   From Audiovisual Speech Enhancement to Audiovisual Source Separation

Introduced first in automatic speech recognition, the bimodality of speech was then exploited to enhance speech signals corrupted by an additive noise [10]. In

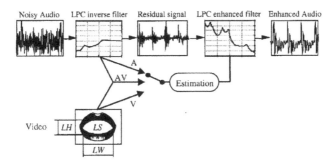

Noisy Audio   LPC inverse filter   Residual signal   LPC enhanced filter   Enhanced Audio

Fig. 2. Audio visual estimation of Wiener filter from [10]

this early study, the system audio speech embedded in white noise is enhanced thanks to a filtering approach, with filter parameters estimated from the video input (see recent developments [6,11,18]). The aim was to estimate Wiener filter

$$H(\nu) = \frac{P_s(\nu)}{P_x(\nu)} \tag{1}$$

where $P_s(\nu)$ and $P_x(\nu)$ are the power spectral densities (PSD) of speech signal $s(t)$ and of noisy signal $x(t)$, respectively. Speech signal $s(t)$ was represented by an autoregressive or linear prediction (LP) model. The visual information was represented by the internal lips height LH, width LW and area LS (Fig. 2). The estimation of Wiener filter was done by a linear regression such that the LP coefficients are obtained from a regressor trained with a training clean database. This approach was shown to be efficient on very simple data (succession of vowels and consonants). However, since the relationship between audio and video signals is complex this approach shows its limits.

In parallel of classical signal enhancement methods, blind source separation (BSS) was a growing field in signal processing. Since BSS is especially suitable to the cocktail party problem, it seems natural that audiovisual source separation became an attractive field.

3.2 Audiovisual Source Separation

One of the first audiovisual source separation algorithms [28,31] is based on the maximisation of the joint audiovisual statistical model $p_{AV}(A(t), V(t))$ between audio feature $A(t)$ and video feature $V(t)$

$$\hat{\mathbf{b}}_1 = \arg\max_{\mathbf{b}_1} \ p_{AV}\left(\mathbf{b}_1^T \mathbf{x}(t), V_1(t)\right), \tag{2}$$

where $\mathbf{x}(t)$ is the mixture vector, and $\mathbf{b}_1$ is the vector used to extract the specific speaker. Due to the ambiguity between audio and video features, the same lip shape could be associated to different sounds[1], it is proposed in [28] to maximise the product of joint audiovisual statistical model over times

$$\hat{\mathbf{b}}_1 = \arg\max_{\mathbf{b}_1} \prod_{t=0}^{T-1} p_{AV}\left(\mathbf{b}_1^T \mathbf{x}(t), V_1(t)\right). \tag{3}$$

One speech source of interest is thus extracted using the visual information simultaneously recorded from the speaker's face by video processing thanks to the audiovisual coherence. However maximising criteria (2) and (3) is computationally expensive especially in the more realistic case of convolutive mixture [36]. As a consequence several methods [22,24,26] based on a combination of audio and audiovisual algorithms were proposed to extract speech sources from convolutive mixture defined by

$$\mathbf{x}(t) = H(t) * \mathbf{s}(t), \tag{4}$$

where the entries of mixing matrix $H(t)$ are the impulse responses of mixing filter $H_{i,j}(t)$, and $*$ denotes the convolution product. Thus in all these studies, the estimation of the separating filter matrix is obtained by an audio only algorithm based on independent component analysis (ICA) applied in the frequency domain where convolutive mixture (4) is transform to several instantaneous problems:

$$\mathbf{X}(t, f) = H(f)\, \mathbf{S}(t, f), \tag{5}$$

where $\mathbf{X}(t, f)$ and $\mathbf{S}(t, f)$ are the short-term Fourier transform of $\mathbf{x}(t)$ and $\mathbf{s}(t)$, respectively, and $H(f)$ is the matrix of frequency response of mixing filter matrix $H(t)$. However, an intrinsic problem of BSS based on ICA in the frequency domain is the permutation indeterminacy [5]. This indeterminacy was first cancelled by maximising the audiovisual coherence between the estimated audio sources and the motion of the speaker's lips thanks to an audiovisual statistical model [23,24]. Nevertheless, this approach is still computationally expensive, and a simpler but still efficient method was proposed [26] to solve the same problem of permutation indeterminacy. It is based on the detection of silent periods of the speaker thanks to a video voice activity detection (V-VAD) [1,29,30]. And lastly, still aiming to simplify the approach the detection of silent period allows to directly extract the corresponding speaker from the mixtures [25]. Indeed, if one source vanishes then the rank of the correlation matrix of mixtures $\mathbf{X}(t, f)$ is $N - 1$ where N is the number of sensors (equal to the number of sources). The eigen vector associated with the null eigen value of this covariance matrix allows to directly estimate the corresponding separation vector. Thus the proposed method [25] consists in silent periods detection for a specific speaker by a V-VAD followed by extracting the speaker signal thanks to the eigenvalue decomposition of the correlation matrix computed within silent periods.

---

[1] For instance the same lip shape is associated with the French sounds [o] and [y].

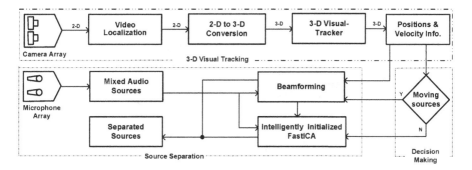

Fig. 3. System block diagram: System block diagram: Video localization is based on state-of-the-art Viola-Jones face detector [35]. The 2-D image information of the two video cameras is converted to 3-D world co-ordinates. The approximated 3-D locations are fed to the tracker, and on the basis of estimated 3-D real world position and velocity from the tracking, the sources are separated either by beamforming or by intelligently initializing the FastICA algorithm [20].

3.3   Intelligent Room

Recently, a significant improvement was made by the application of audiovisual source separation for stationary and moving sources by using the data recorded in the intelligent room.

The System Model. The schematic diagram of the system is shown in Figure 3. The system can be divided into two stages: human tracking to obtain position and velocity information and source separation by utilizing the position and velocity information based on frequency domain BSS algorithms and beamforming.

3-D Tracking and Angle of Arrival Results. Since colour video cameras are used therefore the face detection of the speakers is possible. In Figure 4 the colour blobs indicate that the faces are detected well. The center of the detected face region are approximated as the position of the lips in each sequence at each state.

The approximate 2-D position of the speaker in both synchronized camera frames at each state is converted to 3-D world coordinates. We update the particle filter with this measurement and results of the tracker are shown in Figure 5. To view the tracking results in more detail, the results are plotted in xy axes. Figure 6 clearly shows that the error is almost corrected by the particle filter. The angles of arrivals of moving speakers are shown in Figure 7.

BSS Results. If the sources are moving we separate the sources by the beamformer and when the sources are stationary we separate the sources by intelligently initializing FastICA (IIFastICA). For the stationary case recorded

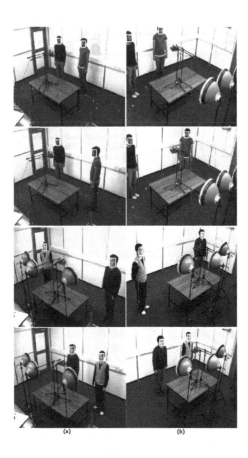

Fig. 4. 3-D Tracking results: frames of synchronized recordings, (a) frames of first camera and (b) frames of second camera; the Viola-Jones face detector [35] efficiently detected the faces in the frames.

mixtures of length of 5sec are separated and results are shown in Figure 8. Length of the data for beamforming case is 0.4sec. The resulting PI [19] in the top of Figures 8 & 9 show good performance i.e. close to zero across the majority of the frequency bins. Permutation was also evaluated based on the criterion mentioned in [19]. In the bottom of Figures 8 & 9 the results confirm that the audio visual information based algorithm automatically mitigates the permutation at each frequency bin. Since there is no permutation problem therefore sources are finally aligned in the time domain. SIR-Improvement [27] for IIFastICA is $12.7dB$ and for beamforming is $9.4dB$. Finally, separation of real room recordings were evaluated subjectively by listening tests, seven people participated in the listening tests and mean opinion score is provided in Table 1.

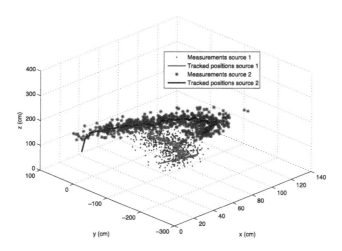

Fig. 5. 3-D Tracking results: PF based 3-D tracking of the speakers while walking around the table in the intelligent office

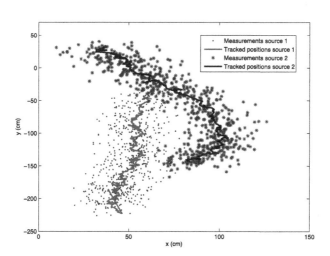

Fig. 6. 3-D Tracking results: PF based tracking of the speakers in the x and y-axis, while walking around the table in the intelligent office. The result provides more in depth view in the x and y-axis.

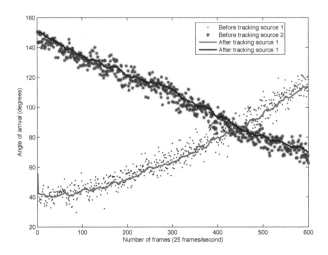

Fig. 7. Angle of arrival results: Angles of arrivals of the speakers to the sensor array. The estimated angles before tracking and corrected angles by PF are shown.

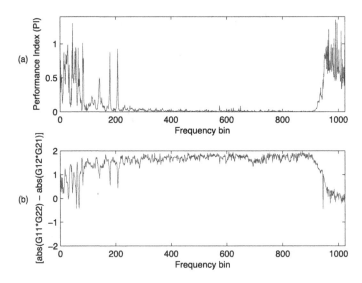

Fig. 8. BSS Results: performance index at each frequency bin for proposed IIFastICA algorithm at the top and evaluation of permutation at the bottom. A lower PI refers to a superior method and $[abs(G_{11}G_{22}) - abs(G_{12}G_{21})] > 0$ means no permutation.

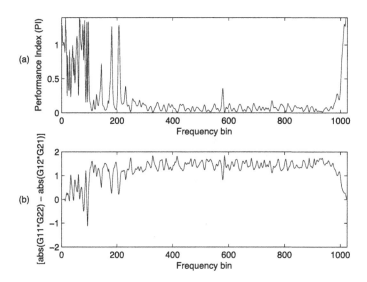

Fig. 9. BSS Results: performance index at each frequency bin for 3-D tracking based angle of arrival information used in beamforming at the top and evaluation of permutation at the bottom. A lower PI refers to a superior method and $[abs(G_{11}G_{22}) - abs(G_{12}G_{21})] > 0$ means no permutation.

Table 1. Subjective evaluation: MOS for separation of real room recordings, by intelligently initialized FastICA (IIFastICA) when sources are stationary, and by beamforming when sources are moving

Algorithms	IIFastICA	Beamforming
Mean opinion score	4.8	3.7

4   Conclusion

Looking at the speakers face can be useful to better hear a speech signal in a noisy environment and to extract it from competing sources. As a consequence, as presented in this brief review of audiovisual extraction of a speaker signal, a lot of methods exploit the coherence between the audio and video modalities, leading thus a efficient extraction algorithms.

References

1. Aubrey, A., Rivet, B., Hicks, Y., Girin, L., Chambers, J., Jutten, C.: Two novel visual voice activity detectors based on appearance models and retinal filltering. In: Proc. European Signal Processing Conference (EUSIPCO), Poznan, Poland, September 2007, pp. 2409–2413 (2007)

2. Benoît, C., Mohamadi, T., Kandel, S.: Effects of phonetic context on audio-visual intelligibility of French. J. Speech and Hearing Research 37, 1195–1293 (1994)
3. Bernstein, L.E., Auer, E.T.J., Takayanagi, S.: Auditory speech detection in noise enhanced by lipreading. Speech Communication 44(1-4), 5–18 (2004)
4. Cherry, E.C.: Some experiments on the recognition of speech, with one and with two ears. Journal of Acoustical Society of America 25(5), 975–979 (1953)
5. Comon, P.: Independent component analysis, a new concept? Signal Processing 36(3), 287–314 (1994)
6. Deligne, S., Potamianos, G., Neti, C.: Audio-Visual speech enhancement with AVCDCN (AudioVisual Codebook Dependent Cepstral Normalization). In: Proc. Int. Conf. Spoken Language Processing (ICSLP), Denver, Colorado, USA, September 2002, pp. 1449–1452 (2002)
7. Ephraim, Y., Malah, D.: Speech Enhancement Using a Minimum Mean-Square Error Short-Time Spectral Amplitude Estimator. IEEE Transactions on Acoustics, Speech, and Signal Processing 32(6), 1109–1121 (1984)
8. Erber, N.P.: Interaction of audition et vision in the recognition of oral speech stimuli. J. Speech and Hearing Research 12, 423–425 (1969)
9. Gannot, S., Burshtein, D., Weinstein, E.: Iterative and sequential kalman filter-based speech enhancement algorithms. IEEE Transactions on Speech and Audio Processing 6(4), 373–385 (1998)
10. Girin, L., Allard, A., Schwartz, J.-L.: Speech signals separation: a new approach exploiting the coherence of audio and visual speech. In: IEEE Int. Workshop on Multimedia Signal Processing (MMSP), Cannes, France (2001)
11. Goecke, R., Potamianos, G., Neti, C.: Noisy audio feature enhancement using audio-visual speech data. In: Proc. IEEE Int. Conf. Acoustics, Speech, and Signal Processing (ICASSP), Orlando, USA, May 2002, pp. 2025–2028 (2002)
12. Grant, K.W., Seitz, P.-F.: The use of visible speech cues for improving auditory detection of spoken sentences. Journal of Acoustical Society of America 108, 1197–1208 (2000)
13. Hérault, J., Jutten, C.: Space or time adaptive signal processing by neural networks models. In: Intern. Conf. on Neural Networks for Computing, Snowbird, USA, pp. 206–211 (1986)
14. Jutten, C., Hérault, J.: Blind separation of sources. Part I: An adaptive algorithm based on a neuromimetic architecture. Signal Processing 24(1), 1–10 (1991)
15. Jutten, C., Taleb, A.: Source separation: from dusk till dawn. In: Proc. Int. Conf. Independent Component Analysis and Blind Source Separation (ICA), Helsinki, Finland, June 2000, pp. 15–26 (2000)
16. Kim, J., Chris, D.: Investigating the audio–visual speech detection advantage. Speech Communication 44(1-4), 19–30 (2004)
17. McGurk, H., McDonald, J.: Hearing lips and seeing voices. Nature 264, 746–748 (1976)
18. Milner, B., Almajai, I.: Noisy audio speech enhancement using wiener filters dervied from visual speech. In: Proc. Int. Conf. Auditory-Visual Speech Processing (AVSP), Moreton Island, Australia (September 2007)
19. Naqvi, S.M., Zhang, Y., Tsalaile, T., Sanei, S., Chambers, J.A.: A multimodal approach for frequency domain independent component analysis with geometrically-based initialization. In: Proc. EUSIPCO, Lausanne, Switzerland (2008)
20. Naqvi, S.M., Zhang, Y., Tsalaile, T., Sanei, S., Chambers, J.A.: A multimodal approach for frequency domain independent component analysis with geometricallybased initialization. In: Proc. EUSIPCO, Lausanne, Switzerland (2008)
21. Potamianos, G., Neti, C., Deligne, S.: Joint Audio-Visual Speech Processing for Recognition and Enhancement. In: Proc. Int. Conf. Auditory-Visual Speech Processing (AVSP), St. Jorioz, France (September 2003)

22. Rivet, B., Girin, L., Jutten, C.: Solving the indeterminations of blind source separation of convolutive speech mixtures. In: Proc. IEEE Int. Conf. Acoustics, Speech, and Signal Processing (ICASSP), Philadelphia, USA, March 2005, pp. V-533–V-536 (2005)
23. Rivet, B., Girin, L., Jutten, C.: Log-Rayleigh distribution: a simple and efficient statistical representation of log-spectral coefficients. IEEE Transactions on Audio, Speech and Language Processing 15(3), 796–802 (2007)
24. Rivet, B., Girin, L., Jutten, C.: Mixing audiovisual speech processing and blind source separation for the extraction of speech signals from convolutive mixtures. IEEE Transactions on Audio, Speech and Language Processing 15(1), 96–108 (2007)
25. Rivet, B., Girin, L., Jutten, C.: Visual voice activity detection as a help for speech source separation from convolutive mixtures. Speech Communication 49(7-8), 667–677 (2007)
26. Rivet, B., Girin, L., Servière, C., Pham, D.-T., Jutten, C.: Using a visual voice activity detector to regularize the permutations in blind source separation of convolutive speech mixtures. In: Proc. Int. Conf. on Digital Signal Processing (DSP), Cardiff, Wales UK, July 2007, pp. 223–226 (2007)
27. Sanei, S., Naqvi, S.M., Chambers, J.A., Hicks, Y.: A geometrically constrained multimodal approach for convolutive blind source separation. In: Proc. IEEE Int. Conf. Acoustics, Speech, and Signal Processing (ICASSP), Honolulu, Hawaii, USA, April 2007, pp. 969–972 (2007)
28. Sodoyer, D., Girin, L., Jutten, C., Schwartz, J.-L.: Developing an audio-visual speech source separation algorithm. Speech Communication 44(1-4), 113–125 (2004)
29. Sodoyer, D., Girin, L., Savariaux, C., Schwartz, J.-L., Rivet, B., Jutten, C.: A study of lip movements during spontaneous dialog and its application to voice activity detection. Journal of Acoustical Society of America 125(2), 1184–1196 (2009)
30. Sodoyer, D., Rivet, B., Girin, L., Schwartz, J.-L., Jutten, C.: An analysis of visual speech information applied to voice activity detection. In: Proc. IEEE Int. Conf. Acoustics, Speech, and Signal Processing (ICASSP), Toulouse, France, pp. 601–604 (2006)
31. Sodoyer, D., Schwartz, J.-L., Girin, L., Klinkisch, J., Jutten, C.: Separation of audio-visual speech sources: a new approach exploiting the audiovisual coherence of speech stimuli. Eurasip Journal on Applied Signal Processing 2002(11), 1165–1173 (2002)
32. Stork, D.G., Hennecke, M.E.: Speechreading by Humans and Machines. Springer, Berlin (1996)
33. Sumby, W., Pollack, I.: Visual contribution to speech intelligibility in noise. Journal of Acoustical Society of America 26, 212–215 (1954)
34. Summerfield, Q.: Some preliminaries to a comprehensive account of audio-visual speech perception. In: Dodd, B., Campbell, R. (eds.) Hearing by Eye: The Psychology of Lipreading, pp. 3–51. Lawrence Erlbaum Associates, Mahwah (1987)
35. Viola, P., Jones, M.: Rapid object detection using a boosted cascade of simple features. In: Proc. IEEE Conf. Comput. Vision Pattern Recognition (CVPR), Kauai, Hawaii, USA, December 2001, pp. 511–518 (2001)
36. Wang, W., Cosker, D., Hicks, Y., Sanei, S., Chambers, J.A.: Video assisted speech source separation. In: Proc. IEEE Int. Conf. Acoustics, Speech, and Signal Processing (ICASSP), Philadelphia, USA (March 2005)

Audio Source Separation Using Hierarchical Phase-Invariant Models

Emmanuel Vincent

INRIA, Centre Inria Rennes - Bretagne Atlantique Campus de Beaulieu,
35042 Rennes Cedex, France
emmanuel.vincent@inria.fr

Abstract. Audio source separation consists of analyzing a given audio recording so as to estimate the signal produced by each sound source for listening or information retrieval purposes. In the last five years, algorithms based on hierarchical phase-invariant models such as single- or multichannel hidden Markov models (HMMs) or nonnegative matrix factorization (NMF) have become popular. In this paper, we provide an overview of these models and discuss their advantages compared to established algorithms such as nongaussianity-based frequency-domain independent component analysis (FDICA) and sparse component analysis (SCA) for the separation of complex mixtures involving many sources or reverberation. We argue how hierarchical phase-invariant modeling could form the basis of future modular source separation systems.

1 Introduction

Most audio signals are mixtures of several sound sources which are active simultaneously. For example, speech recordings in "cocktail party" environments are mixtures of several speakers, music CDs are mixtures of musical instruments and singers, and movie soundtracks are mixtures of speech, music and environmental sounds. Audio source separation is the problem of recovering the individual source signals underlying a given mixture.

Two alternative approaches to this problem have emerged: computational auditory scene analysis (CASA) and Bayesian inference. CASA consist of building auditory-motivated sound processing systems composed of four successive modules: front-end auditory representation, low-level primitive grouping, higher-level schema-based grouping and binary time-frequency masking. By contrast, the Bayesian approach consists of building probabilistic generative models of the source signals and estimating them in a minimum mean squared error (MMSE) or maximum a posteriori (MAP) sense from the mixture. The generative models are defined via latent variables and prior conditional distributions between variables. Although individual CASA modules are sometimes amenable to probabilistic models and inference criteria, the Bayesian approach is potentially more robust since all available priors are jointly taken into account via top-down feedback.

J. Solé-Casals and V. Zaiats (Eds.): NOLISP 2009, LNAI 5933, pp. 12–16, 2010.

Most established Bayesian source separation algorithms rely on time-frequency domain linear modeling [1]. Assuming point sources and low reverberation, the mixing process can be approximated as linear time-invariant filtering. The vector $\mathbf{X}_{nf}$ of complex-valued short-time Fourier transform (STFT) coefficients of all channels of the mixture signal in time frame n and frequency bin f is given by

$$\mathbf{X}_{nf} = \sum_{j=1}^{J} S_{jnf}\mathbf{A}_{jf} + \mathbf{E}_{nf} \tag{1}$$

where S_{jnf} are the scalar STFT coefficients of the J underlying single-channel source signals indexed by j, $\mathbf{A}_{jf}$ are mixing vectors representing the frequency response of the mixing filters and $\mathbf{E}_{nf}$ is some residual noise. The mixing vectors are typically modeled conditionally to the source directions of arrival via instantaneous or near-anechoic priors, while the source STFT coefficients are modeled as independent and identically distributed according to binary or continuous sparse priors. These priors yield different classes of source separation algorithms, including spatial time-frequency masking, nongaussianity-based frequency-domain independent component analysis (FDICA) and sparse component analysis (SCA) [1].

While these algorithms have achieved astounding results on certain mixtures, their performance significantly degrades on complex sound scenes involving many sources or reverberation [2]. Indeed, due to use of low-informative source priors, separation relies mostly on spatial cues, which are obscured in complex situations. In order to address this issue, additional spectral cues must be exploited. In the framework of linear modeling, this translates into parameterizing each source signal as a linear combination of sound atoms representing for instance individual phonemes or musical notes. In theory, a huge number of sound atoms is needed to obtain an accurate representation since most sources produce phase-invariant atoms characterized by stable variance but somewhat random phase at each frequency. In practice however, only a relatively small number of atoms is usually assumed due to computational constraints, resulting in limited performance improvement [3].

2 Hierarchical Phase-Invariant Models

2.1 General Formulation

Explicit phase invariance can be ensured instead by modeling the source STFT coefficients in a hierarchical fashion via a non-sparse circular distribution whose parameters vary over the time-frequency plane according to some prior. This model appears well suited to audio signals, which are typically non-sparse over small time-frequency regions but non-stationary hence sparse over larger regions. Different distributions have been investigated. Assuming that the source STFT coefficients follow a zero-mean Gaussian distribution, the vector $\mathbf{X}_{nf}$ of mixture STFT coefficients in time-frequency bin (n, f) obeys the zero-mean multivariate Gaussian model [4,5]

$$\mathbf{X}_{nf} \sim \mathcal{N}\left(\mathbf{0}, \sum_{j=1}^{J} V_{jnf}\mathbf{R}_{jf}\right) \qquad (2)$$

where V_{jnf} are the scalar variances or power spectra of the sources and $\mathbf{R}_{jf}$ are Hermitian mixing covariance matrices. In the particular case of a stereo mixture, each of these covariance matrices encodes three spatial quantities: interchannel intensity difference (IID), interchannel phase difference (IPD) and interchannel correlation or coherence (IC) [6]. Multichannel log-Gaussian distributions based on these quantities and single-channel log-Gaussian and Poisson distributions have also been proposed [6,7,8].

2.2   Prior Distributions over the Model Parameters

Three nested families of prior distributions over the variance parameters V_{jnf} have been explored so far. In [9,4,5,10,11], the variance of each source is assumed to be locally constant over small time-frequency regions and uniformly or sparsely distributed. In [7,12,13,14], the spectro-temporal distribution of variance is constrained by a Gaussian mixture model (GMM) or, more generally, a hidden Markov model (HMM) that describe each source on each time frame by a latent discrete state indexing one of a set of template spectra. In [15,6,8,16], the spectro-temporal distribution of variance is modeled on each time frame by a linear combination of basis spectra weighted by continuous latent scaling factors. The template spectra and the basis spectra may be either learned on specific training data for each source [7,15,6], or learned on the same set of training data for all sources [12,14] or adapted to the mixture [13,8].

Assuming point sources and low reverberation, the mixing covariance matrices have rank 1 and can be modeled conditionally to the source directions using instantaneous or near-anechoic priors over the aforementioned linear mixing vectors $\mathbf{A}_{jf}$ as $\mathbf{R}_{jf} = \mathbf{A}_{jf}\mathbf{A}_{jf}^{H}$ [9,4]. The model extends to diffuse sources or reverberant conditions, that translate into full-rank mixing covariance matrices. Full-rank uniform priors have been considered in [5,11].

2.3   Inference Algorithms and Results

Approximate inference for the above model is generally carried out by first estimating the model parameters in the MAP sense then deriving the MMSE source STFT coefficients by Wiener filtering. Depending on the chosen priors, different classes of algorithms may be employed to estimate the model parameters, including nonstationarity-based FDICA and SCA [9,10,11], expectation-maximization (EM) decoding of GMM and HMM [7] and nonnegative matrix factorization (NMF) [8]. Although it is not always the most efficient, the EM algorithm is easily applicable in all cases involving a Gaussian distribution.

For single-channel mixtures, the reported signal-to-distortion ratios (SDRs) are on the order of 7 decibels (dB) on mixtures of two speech sources [8] and 10 dB on mixtures of singing voice and musical accompaniment [13]. For stereo mixtures, nonstationarity-based FDICA and SCA have been shown to outperform nongaussianity-based FDICA and SCA by 1 dB or more [17,10,11]. Multichannel NMF has even improved the SDR by 10 dB or more compared to nongaussianity-based FDICA and SCA on certain very reverberant music mixtures with known source directions and learned instrument-specific basis spectra [6].

3 Conclusion

To conclude, we believe that hierarchical phase-invariant modeling is a promising framework for research into high-quality source separation. Indeed it allows at the same time accurate modeling of diffuse or reverberant sources and efficient exploitation of spectral cues. These advantages are essential for accurate source discrimination in complex mixtures, involving many sources or strong reverberation. Yet, the potential modeling capabilities offered by this framework have only been little explored. On the one hand, existing models involve a large number of latent variables encoding low-level information. Separation performance and robustness could increase by conditioning these variables on additional latent variables encoding higher-level information, such as reverberation time, source directivity, voiced/unvoiced character and fundamental frequency. On the other hand, most existing source separation systems rely on a single class of priors, which may not be optimal for all sources in a real-world scenario. This limitation could be addressed by designing modular systems combining possibly different classes of priors for each source, selected either manually or automatically. Recent studies in these two directions have introduced promising ideas for future research in this area [18,11,19].

References

1. Makino, S., Lee, T.W., Sawada, H. (eds.): Blind speech separation. Springer, Heidelberg (2007)
2. Vincent, E., Araki, S., Bofill, P.: The 2008 Signal Separation Evaluation Campaign: A community-based approach to large-scale evaluation. In: Proc. 8th Int. Conf. on Independent Component Analysis and Signal Separation, pp. 734–741 (2009)
3. Gowreesunker, B.V., Tewfik, A.H.: Two improved sparse decomposition methods for blind source separation. In: Proc. 7th Int. Conf. on Independent Component Analysis and Signal Separation, pp. 365–372 (2007)
4. Févotte, C., Cardoso, J.F.: Maximum likelihood approach for blind audio source separation using time-frequency Gaussian models. In: Proc. 2005 IEEE Workshop on Applications of Signal Processing to Audio and Acoustics, pp. 78–81 (2005)
5. El Chami, Z., Pham, D.T., Servière, C., Guerin, A.: A new model-based underdetermined source separation. In: Proc. 11th Int. Workshop on Acoustic Echo and Noise Control (2008); paper ID 9061
6. Vincent, E.: Musical source separation using time-frequency source priors. IEEE Trans. Audio Speech Lang. Process. 14, 91–98 (2006)
7. Roweis, S.T.: One microphone source separation. Advances in Neural Information Processing Systems 13, 793–799 (2001)

8. Virtanen, T., Cemgil, A.T.: Mixtures of gamma priors for non-negative matrix factorization based speech separation. In: Proc. 8th Int. Conf. on Independent Component Analysis and Signal Separation, pp. 646–653 (2009)
9. Pham, D.T., Servière, C., Boumaraf, H.: Blind separation of speech mixtures based on nonstationarity. In: Proc. 7th Int. Symp. on Signal Processing and its Applications, pp. II-73–II-76 (2003)
10. Vincent, E., Arberet, S., Gribonval, R.: Underdetermined instantaneous audio source separation via local Gaussian modeling. In: Proc. 8th Int. Conf. on Independent Component Analysis and Signal Separation, pp. 775–782 (2009)
11. Duong, N.Q.K., Vincent, E., Gribonval, R.: Spatial covariance models for underdetermined reverberant audio source separation. In: Proceedings of the 2009 IEEE Workshop on Applications of Signal Processing to Audio and Acoustics, pp. 129–132 (2009)
12. Attias, H.: New EM algorithms for source separation and deconvolution with a microphone array. In: Proc. 2003 IEEE Int. Conf. on Acoustics, Speech and Signal Processing, pp. V-297–V-300 (2003)
13. Ozerov, A., Philippe, P., Bimbot, F., Gribonval, R.: Adaptation of Bayesian models for single-channel source separation and its application to voice/music separation in popular songs. IEEE Trans. Audio Speech Lang. Process. 15, 1564–1578 (2007)
14. Nix, J., Hohmann, V.: Combined estimation of spectral envelopes and sound source direction of concurrent voices by multidimensional statistical filtering. IEEE Trans. Audio Speech Lang. Process. 15, 995–1008 (2007)
15. Benaroya, L., McDonagh, L., Bimbot, F., Gribonval, R.: Non negative sparse representation for Wiener based source separation with a single sensor. In: Proc. 2003 IEEE Int. Conf. on Acoustics, Speech and Signal Processing, pp. VI-613–VI-616 (2003)
16. Ozerov, A., Févotte, C.: Multichannel nonnegative matrix factorization in convolutive mixtures. With application to blind audio source separation. In: Proc. 2009 IEEE Int. Conf. on Acoustics, Speech and Signal Processing, pp. 3137–3140 (2009)
17. Puigt, M., Vincent, E., Deville, Y.: Validity of the independence assumption for the separation of instantaneous and convolutive mixtures of speech and music sources. In: Proc. 8th Int. Conf. on Independent Component Analysis and Signal Separation, pp. 613–620 (2009)
18. FitzGerald, D., Cranitch, M., Coyle, E.: Extended nonnegative tensor factorisation models for musical sound source separation. Computational Intelligence and Neuroscience, article ID 872425 (2008)
19. Blouet, R., Rapaport, G., Cohen, I., Févotte, C.: Evaluation of several strategies for single sensor speech/music separation. In: Proc. 2008 IEEE Int. Conf. on Acoustics, Speech and Signal Processing, pp. 37–40 (2008)

Visual Cortex Performs a Sort of Non-linear ICA

Jesús Malo and Valero Laparra

Image Processing Laboratory, Universitat de València.
Dr. Moliner 50, 46100 Burjassot, València, (Spain).
Jesus.Malo@uv.es, Valero.Laparra@uv.es
http://www.uv.es/vista/vistavalencia

Abstract. Here, the standard V1 cortex model optimized to reproduce image distortion psychophysics is shown to have nice statistical properties, e.g. approximate factorization of the PDF of natural images. These results confirm the *efficient encoding hypothesis* that aims to explain the organization of biological sensors by information theory arguments.[1]

1 Introduction

More than fifty years ago it was suggested that image data in the human visual system (HVS) should be represented by using components as independent from each other as posible (Barlow, 1961). This was called thereafter as the *efficient encoding hypothesis*. The linear processing in the HVS follows this interpretation since the shape of the linear receptive fields can be predicted by optimizing a set of linear filters using linear independent components analysis (ICA) over natural images (Olshausen and Field, 1996). However, although *linear* ICA basis functions are optimized to reduce the mutual information (redundancy reduction goal), they can not remove completely the redundancy (Simoncelli and Olshausen, 2001). Therefore a *non-linear* step is needed in order to obtain a representation in which the components are independent from each other.

In this work we show how the standard HVS model with non-linear gain control approximately factorizes the PDF of natural images. Moreover the capability of this model for redundancy reduction is tested on natural images by computing the mutual information reduction obtained by the linear stage and the non-linear stage. Since the model is fitted psychophysically and no statistical information is used, the results suggest that the HVS performs a sort of non-linear ICA over the natural images.

The structure of the paper is as follows. In section 2 we review the standard non-linear model of the V1 visual cortex and propose a new method to set its parameters. Section 3 analytically shows how the proposed model may factorize a plausible PDF for natural images. Section 4 empirically shows how the

---

[1] This work was partially supported by projects CICYT-FEDER TEC2009-13696 and CSD2007-00018, and grant BES2007-16125.

J. Solé-Casals and V. Zaiats (Eds.): NOLISP 2009, LNAI 5933, pp. 17–25, 2010.

proposed model achieves component independence and redundancy reduction. Finally, section 5 draws the conclusions of the work.

2   V1 Visual Cortex Model

The image representation considered here is based on the standard psychophysical and physiological model that describes the early visual processing up to the V1 cortex (Mullen, 1985; Malo, 1997; Heeger, 1992; Watson and Solomon, 1997). In this model, the input image, $\mathbf{x} = (x_1, \cdots, x_n)$, is first analyzed by a set of wavelet-like linear sensors, $\mathbf{T}_{ij}$, that provide a scale and orientation decomposition of the image (Watson and Solomon, 1997). The linear sensors have a frequency dependent linear gain according to the Contrast Sensitivity Function (CSF), $\mathbf{S}_i$, (Mullen, 1985; Malo, 1997). The weighted response of these sensors is non-linearly transformed according to the Divisive Normalization gain control, $\mathbf{R}$ (Heeger, 1992; Watson and Solomon, 1997):

$$\mathbf{x} \xrightarrow{\mathbf{T}} \mathbf{w} \xrightarrow{\mathbf{S}} \mathbf{w}' \xrightarrow{\mathbf{R}} \mathbf{r} \tag{1}$$

In this scheme, the rows of the matrix $\mathbf{T}$ contain the receptive fields of V1 neurons, here modeled by an orthogonal 4-scales QMF wavelet transform . $\mathbf{S}$ is a diagonal matrix containing the linear gains to model the CSF. Finally, $\mathbf{R}$ is the Divisive Normalization response:

$$\mathbf{R}(\mathbf{w}')_i = r_i = \operatorname{sign}(w_i')\frac{|S_i \cdot w_i|^\gamma}{\beta_i^\gamma + \sum_{k=1}^n H_{ik}|S_k \cdot w_k|^\gamma} \tag{2}$$

where H is a kernel matrix that controls how the responses of neighboring linear sensors, k, affect the non-linear response of sensor i. Here we use the Gaussian interaction kernel proposed by Watson and Solomon (Watson and Solomon, 1997), which has been successfully used in block-frequency domains (Malo et al., 2006; Gutiérrez et al., 2006; Camps et al., 2008). In the wavelet domain the width of the interaction kernel for spatial, orientation and scale neighbors has to be found. The resulting kernel is normalized to ensure that $\sum_k H_{ik} = 1$. In our implementation of the model we set the profile of the regularizing constants β_i according to the standard deviation of each subband of the wavelet coefficients of natural images in the selected wavelet representation. This initial guess is consistent with the interpretation of the values β_i as priors of the amplitude of the coefficients (Schwartz and Simoncelli, 2001). This profile (computed from 100 images of a calibrated image data base (http://tabby.vision.mcgill.ca) is further multiplied by a constant to be fitted to the psychophysical data.

The above V1 image representation induces a subjective image distortion metric. Given an input image, $\mathbf{x}$, and its distorted version, $\mathbf{x}' = \mathbf{x} + \Delta\mathbf{x}$, the model provides two response vectors, $\mathbf{r}$, and $\mathbf{r}' = \mathbf{r} + \Delta\mathbf{r}$. The perceived distortion has been proposed to be the Euclidean norm of the difference vector (Teo and Heeger, 1994), but non-quadratic pooling norms have also been reported (Watson and Solomon, 1997; Watson and Malo, 2002).

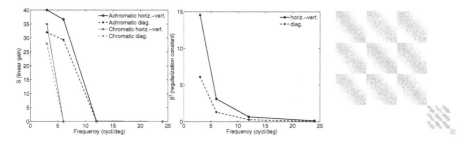

Fig. 1. Linear gains S (left), regularization constants β^γ (center), and kernel H (right)

The color version of the V1 response model involves the same kind of spatial transforms described above applied on the image channels in an opponent color space (Martinez-Uriegas, 1997). According to the well known differences in frequency sensitivity in the opponent channels (Mullen, 1985), we will allow for different matrices **S** in each channel. We will assume the same behavior for the other spatial transforms since the non-linear behavior of the chromatic channels is similar to the achromatic non-linearities (Martinez-Uriegas, 1997).

The natural way to set the parameters of the model is by fitting threshold psychophysics or physiological recordings (Heeger, 1992) (Watson and Solomon, 1997). This low-level approach is not straightforward because the experimental literature is often interested in a subset of the parameters, and a variety of experimental settings is used. As a result, it is not easy to unify the wide range of data into a common computational framework. Alternative (theoretical) approaches involve using image statistics and the efficient encoding hypothesis(Olshausen and Field, 1996; Schwartz and Simoncelli, 2001) (Malo and Gutiérrez, 2006), but that is not the right thing to do since we want to include no statistical information in the model.

Instead, in this work we used an empirical but *indirect* approach: we set the parameters of the model to reproduce experimental (but higher-level) visual results such as image quality assessment as in (Watson and Malo, 2002). In particular, we optimized the Divisive Normalization metric to maximize the correlation with the subjective ratings of a subset of the LIVE Quality Assessment Database[2]. The range of the parameter space was set according to an initial guess obtained from threshold psychophysics (Mullen, 1985; Watson and Solomon, 1997; Malo, 1997) and previous use of similar models in image processing applications (Malo et al., 2006; Gutiérrez et al., 2006; Camps et al., 2008).

Figure 1 shows the optimal values for the linear gains **S**, the regularization constants β^γ and the interaction kernel H. The particular structure of the interaction kernel comes from the particular arrangement of wavelet coefficients used in the transform. The optimal value for the excitation and inhibition exponent was $\gamma = 1.7$. The optimal values for the spatial and frequency summation exponents were $q_s = 3.5$ and $q_f = 2$, where the summation is made first over space and then over the frequency dimensions.

---

[2] http://live.ece.utexas.edu/research/quality/

3 PDF Factorization through V1 Divisive Normalization

In this section we assume a plausible joint PDF model for natural images in the wavelet domain and we show that this PDF is approximately factorized by a divisive normalization transform, given that some conditions apply. The analytical results shown here predict quite characteristic marginal PDFs in the transformed domain. In section 4 we will empirically check the predictions made here by applying the model proposed above to a set of natural images.

3.1 Image Model

It is widely known that natural images display a quite characteristic behavior in the wavelet domain: on the one hand, they show heavy-tailed marginal PDFs, $P_{w'_i}(w'_i)$ (see Fig. 2), and, on the other hand, the variance of one particular coefficient is related to the variance of the neighbors. This quite evident by looking at the so called bow-tie plot: the conditional probability of a coefficient given the values of some of its neighbors, $P(w'_i|w'_j)$, normalized by the maximum of the function for each value of w'_j (see Fig. 2). These facts have been used to propose leptokurtotic functions to model the marginal PDFs (Hyvärinen, 1999) and models of the conditional PDFs in which the variance of one coefficient depends on the variance of the neighbors (Schwartz and Simoncelli, 2001).

Inspired on these conditional models, we propose the following joint PDF (for the N-dimensional vectors $\mathbf{w'}$), in which, each element of the diagonal covariance, Σ_{ii}, depends on the neighbors:

$$P_{\mathbf{w'}}(\mathbf{w'}) = \mathcal{N}(0, \Sigma(\mathbf{w'})) = \frac{1}{(2\pi)^{N/2}|\Sigma(\mathbf{w'})|^{1/2}} \, e^{-\frac{1}{2}\mathbf{w'}^T \cdot \Sigma(\mathbf{w'})^{-1} \cdot \mathbf{w'}} \tag{3}$$

where,

$$\Sigma_{ii}(\mathbf{w'}) = (\beta_i^\gamma + \sum_j H_{ij} \cdot |w'_j|^\gamma)^{\frac{2}{\gamma}} \tag{4}$$

Note that this joint PDF *is not* Gaussian because the variance of each coefficient depends on the neighbors according to the kernel in eq. 4. Therefore, the coefficients of the wavelet transform are not independent since the joint PDF, $P_{\mathbf{w'}}(\mathbf{w'})$, cannot be factorized by its marginal PDFs, $P_{w'_i}(w'_i)$.

A 2D toy example using using the above joint PDF illustrates its suitability to capture the reported marginal and conditional behavior of wavelet coefficients: see the predictions shown in Fig. 2).

3.2 V1 Normalized Components Are Approximately Independent

Here we compute the PDF of the natural images in the divisive normalized representation assuming (1) the above image model, and (2) the match between the denominator of the normalization and the covariance of the image model.

We will use the fact that given the PDF of a random variable, $\mathbf{w}'$, and some transform, $\mathbf{r} = \mathbf{R}(\mathbf{w}')$, the PDF of the transformed variable can be computed by (Stark and Woods, 1994),

$$P_{\mathbf{r}}(\mathbf{r}) = P_{\mathbf{w}'}(\mathbf{R}^{-1}(\mathbf{r})) \cdot |\nabla_{\mathbf{r}}\mathbf{R}^{-1}|$$

Considering that the divisive normalization (in vector notation) is just: $\mathbf{r} = sign(\mathbf{w}') \, \Sigma(\mathbf{w}')^{-\frac{\gamma}{2}} \cdot |\mathbf{w}'|^{\gamma}$, where $|\cdot|^{\gamma}$ is an element-by-element exponentiation, the inverse, $\mathbf{R}^{-1}$, can be obtained from one of these (equivalent) expressions (Malo et al., 2006):

$$|\mathbf{w}'|^{\gamma} = (I - D_{|\mathbf{r}|}H)^{-1} \cdot D_{\beta^{\gamma}} \cdot |\mathbf{r}| \tag{5}$$

$$\mathbf{w}' = sign(\mathbf{r})\Sigma(\mathbf{w}')^{\frac{1}{2}} \cdot |\mathbf{r}|^{\frac{1}{\gamma}} \tag{6}$$

where D_v are diagonal matrices with the vector v in the diagonal. Plugging $\mathbf{w}'$ into the image model and using $|\mathbf{w}'|^{\gamma}$ to compute the Jacobian of the inverse,

$$P_{\mathbf{w}'}(\mathbf{R}^{-1}(\mathbf{r})) = \frac{1}{(2\pi)^{N/2} \, |\Sigma(\mathbf{w}')|^{1/2}} \, e^{-\frac{1}{2}(\mathbf{r}^{1/\gamma})^{T} \cdot I \cdot (\mathbf{r}^{1/\gamma})}$$

$$|\nabla_{\mathbf{r}}\mathbf{R}^{-1}| = det\left(\frac{1}{\gamma}\Sigma(\mathbf{w})^{1/2} \cdot D_{|\mathbf{r}|^{\frac{1}{\gamma}-1}} \cdot \left(I + \underbrace{D_{\beta^{-1}} \cdot H \cdot (I - D_{|\mathbf{r}|}H)^{-1} \cdot D_{\beta} \cdot D_{|\mathbf{r}|}}\right)\right)$$

Assuming that the matrix in the brace is negligible

$$|\nabla_{\mathbf{r}}\mathbf{R}^{-1}| \sim det(\Sigma(\mathbf{w}'))^{1/2} \prod_{i=1}^{N} \frac{1}{\gamma} r_i^{\frac{1}{\gamma}-1} \tag{7}$$

it follows that the joint PDF of the normalized signal is just the product of N functions that depend solely on r_i:

$$P_{\mathbf{r}}(\mathbf{r}) = \prod_{i=1}^{N} \frac{1}{\gamma(2\pi)^{1/2}} \, r_i^{\frac{1}{\gamma}-1} \, e^{-\frac{r_i^{2/\gamma}}{2}} = \prod_{i=1}^{N} P_{r_i}(r_i) \tag{8}$$

i.e., we have factorized the joint PDF into its marginal PDFs.

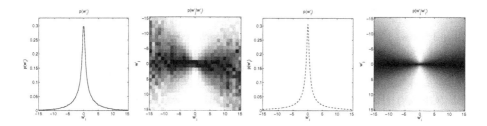

Fig. 2. Left: empirical behavior of wavelet coefficients of natural images (marginal PDF and conditional PDF). Right: simulated behavior according to the proposed model. In this toy experiment we considered two coefficients of the second scale of $\mathbf{w}'$ (computed for 8000 images). We used $S_i = 0.14$, $\beta_i = 0.4$, $H_{ii} = 0.7$ and $H_{ij} = 0.3$ and $\gamma = 1.7$, according to the psychophysically fitted model.

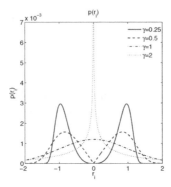

Fig. 3. Family of marginal PDFs of the normalized coefficients r_i as a function of γ

Even though factorization of the PDF does not depend on γ, it determines the shape of the marginal PDFs (see Fig. 3). However, note that different values of γ would imply a better (or worse) match between the denominator of the normalization and the covariance of the image model.

4   Results

This section assesses the component independence performance of the psychophysically fitted V1 image representation (i.e. the validity of Eq. 3) by (1) Mutual Information (MI) measures, and (2) by analyzing the conditional probabilities of the transformed coefficients. To do so, 8000 image patches, **x**, of size 72×72 were considered and transformed to the linear wavelet domain, **w**, and to the non-linear V1 representation, **r**. For the sake of illustration, the results for two values of the exponent γ are used in the divisive normalization: the psychophysically optimal value $\gamma = 1.7$, and $\gamma = 0.5$ due to the (predicted) characteristic shape of the marginal PDFs in that case (see Fig. 3).

4.1   Mutual Information Measures

Table 1 shows the MI results (in bits) for pairs of coefficients in **w** and **r**. 120000 pairs of coefficients were used in each estimation. Two kinds of MI estimators were used: (1) direct computation of MI, which involves 2D histogram estimation (Cover and Tomas, 1991), and (2) estimation of MI by PCA-based Gaussianization (GPCA) (Laparra et al., 2009), which only involves univariate histogram estimations.

These results show that the wavelet representation removes about 92% of the redundancy in the spatial domain, and divisive normalization further reduces about 69% of the remaining redundancy. This suggests that one of the goals of the psychophysical V1 image representation is redundancy reduction.

Table 1. MI measures in bits. GPCA MI estimations are shown in parenthesis. Just for reference, the MI among luminance values in the spatial domain is 2.12 (2.14) bits.

	w	$r^{(0.5)}$	$r^{(1.7)}$
Intraband (scale = 2)	0.29 (0.27)	0.17 (0.17)	0.16 (0.15)
Intraband (scale = 3)	0.24 (0.22)	0.08 (0.09)	0.09 (0.09)
Inter-scale, scales = (1,2)	0.17 (0.17)	0.10 (0.11)	0.08 (0.08)
Inter-scale, scales = (2,3)	0.17 (0.15)	0.04 (0.04)	0.04 (0.04)
Inter-scale, scales = (3,4)	0.09 (0.07)	0.01 (0.01)	0.01 (0.01)
Inter-orientation (H-V), scale = 2	0.10 (0.08)	0.01 (0.01)	0.01 (0.01)
Inter-orientation (H-V), scale = 3	0.08 (0.06)	0.01 (0.01)	0.01 (0.01)
Inter-orientation (H-D), scale = 2	0.16 (0.15)	0.04 (0.04)	0.03 (0.03)
Inter-orientation (H-D), scale = 3	0.15 (0.14)	0.01 (0.01)	0.02 (0.02)

4.2   Marginal and Conditional PDFs

Figure 4 shows the predicted and the experimental marginal PDFs in the normalized domain and the experimental conditional PDFs. The resemblance among theory and experiments confirms the theoretical results in section 3. Note also that the PDF of one coefficient given the neighbor is more independent of the

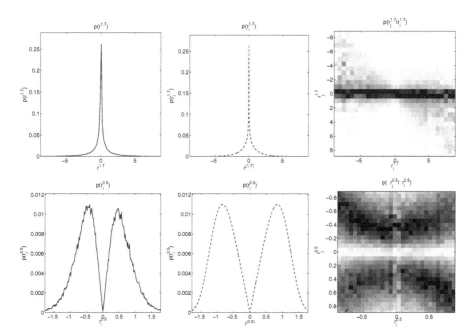

Fig. 4. Experimental marginal PDF (left), theoretical prediction (center), and bowtie plots (right) for **r** coefficients using the optimal value of $\gamma = 1.7$ (top) and other illustrative value, $\gamma = 0.5$ (bottom)

neighbor value than in the wavelet domain (Fig. 2). This is particularly true in the case of using the optimal value $\gamma = 1.7$, thus indicating the match of the physchophysically optimal vision model and image statistics. Note also that the agreement between the marginal PDFs and the theoretical prediction is better for the optimal exponent.

5   Conclusions

Here we showed that the standard V1 cortex model optimized to reproduce image quality psychophysics increases the independence of the image coefficients obtained by linear ICA (wavelet-like) filters. Theoretical results (confirmed by experiments) show that the V1 model approximately factorizes a plausible joint PDF in the wavelet domain: bow-tie dependencies are almost removed and redundancy is substantially reduced.

The results presented here strongly suggest that the early stages of visual processing (up to V1) perform a sort of non-linear ICA on the input images.

References

[Barlow, 1961]Barlow, H.: Possible principles underlying the transformation of sensory messages. In: Rosenblith, W. (ed.) Sensory Communication, pp. 217–234. MIT Press, Cambridge (1961)

[Camps et al., 2008]Camps, G., Gutiérrez, J., Gómez, G., Malo, J.: On the suitable domain for SVM training in image coding. JMLR 9, 49–66 (2008)

[Cover and Tomas, 1991]Cover, T., Tomas, J.: Elements of Information Theory. John Wiley & Sons, New York (1991)

[Gutiérrez et al., 2006]Gutiérrez, J., Ferri, F., Malo, J.: Regularization operators for natural images based on nonlinear perception models. IEEE Tr. Im. Proc. 15(1), 189–200 (2006)

[Heeger, 1992]Heeger, D.J.: Normalization of cell responses in cat striate cortex. Visual Neuroscience 9, 181–198 (1992)

[Hyvärinen, 1999]Hyvärinen, A.: Sparse code shrinkage: Denoising of nongaussian data by ML estimation. Neur. Comp., 1739–1768 (1999)

[Laparra et al., 2009]Laparra, V., Camps, G., Malo, J.: PCA gaussianization for image processing. In: Proc. IEEE ICIP 2009, pp. 3985–3988 (2009)

[Malo, 1997]Malo, J.: Characterization of HVS threshold performance by a weighting function in the Gabor domain. J. Mod. Opt. 44(1), 127–148 (1997)

[Malo et al., 2006]Malo, J., Epifanio, I., Navarro, R., Simoncelli, E.: Non-linear image representation for efficient perceptual coding. IEEE Transactions on Image Processing 15(1), 68–80 (2006)

[Malo and Gutiérrez, 2006]Malo, J., Gutiérrez, J.: V1 non-linear properties emerge from local-to-global non-linear ICA. Network: Computation in Neural Systems 17, 85–102 (2006)

[Martinez-Uriegas, 1997]Martinez-Uriegas, E.: Color detection and color contrast discrimination thresholds. In: Proc. OSA Meeting, p. 81 (1997)

[Mullen, 1985]Mullen, K.T.: The CSF of human colour vision to red-green and yellow-blue chromatic gratings. J. Physiol. 359, 381–400 (1985)

[Olshausen and Field, 1996]Olshausen, B.A., Field, D.J.: Emergence of simple-cell receptive field properties by learning a sparse code for natural images. Nature 381, 607–609 (1996)

[Schwartz and Simoncelli, 2001]Schwartz, O., Simoncelli, E.: Natural signal statistics and sensory gain control. Nat. Neurosci. 4(8), 819–825 (2001)

[Simoncelli and Olshausen, 2001]Simoncelli, E., Olshausen, B.: Natural image statistics and neural representation. Annu. Rev. Neurosci. 24, 1193–1216 (2001)

[Stark and Woods, 1994]Stark, H., Woods, J.: Probability, Random Processes, and Estimation Theory for Engineers. Prentice Hall, NJ (1994)

[Teo and Heeger, 1994]Teo, P., Heeger, D.: Perceptual image distortion. In: Proceedings of the SPIE, vol. 2179, pp. 127–141 (1994)

[Watson and Malo, 2002]Watson, A., Malo, J.: Video quality measures based on the standard spatial observer. In: Proc. IEEE ICIP, vol. 3, pp. 41–44 (2002)

[Watson and Solomon, 1997]Watson, A., Solomon, J.: A model of visual contrast gain control and pattern masking. JOSA A 14, 2379–2391 (1997)

High Quality Emotional HMM-Based Synthesis in Spanish

Xavi Gonzalvo[1,2], Paul Taylor[1], Carlos Monzo[2], Ignasi Iriondo[2],
and Joan Claudi Socoró[2]

[1] Phonetic-Arts Ltd.
St. John's Innovation Center, Cambridge, UK
[2] Enginyeria i Arquitectura La Salle, Universitat Ramon Llull
Grup de Recerca en Processament
{xavi.gonzalvo,paul.taylor}@phonetic-arts.com,
{gonzalvo,cmonzo,iriondo,jclaudi}@salle.url.edu

Abstract. This paper describes a high-quality Spanish HMM-based speech synthesis of emotional speaking styles. The quality of the HMM-based speech synthesis is enhanced by using the most recent features presented for the Blizzard system (i.e. STRAIGHT spectrum extraction and mixed excitation). Two techniques are evaluated. First, a method simultaneously model all emotions within a single acoustic model. Second, an adaptation techniques to convert a neutral emotional style to a target emotion. We consider 3 kinds of emotions expressions: neutral, happy and sad. A subjective evaluation will show the quality of the system and the intensity of the produced emotion while an objective evaluation based on voice quality parameters evaluates the effectiveness of the approaches.

Keywords: HMM-based speech synthesis, emotion, adaptation.

1 Introduction

Current state-of-the-art text-to-speech (TTS) systems are often able to produce intelligible and natural speech, but the speaking style is typically restricted to neutral style and so does not exhibit the full range of natural speech. Hence the development of more expressive and natural speech synthesis systems is becoming an important research area. Approaches to this reflect the two main synthesis techniques today. One approach is to use concatenative speech synthesis, which produces good quality speech but requires a large amount of data for each emotion [1] [2]. On the other hand, we can use a modelling approach and transform the synthesized speech into a target emotion by modelling changes in prosody and voice quality (VoQ) [3]. A significant recent development in hidden Markov model (HMM) synthesis has been to create a speaker independent voice, trained from multiple speakers, that can be later be transformed to a single speaker's voice by using speaker adaptation techniques on a small amount of data [4]. These same techniques can also be used to adapt to a target emotion style [5].

J. Solé-Casals and V. Zaiats (Eds.): NOLISP 2009, LNAI 5933, pp. 26–34, 2010.

In order to reproduce a speaking style for a specific emotion and maintain naturalness, it is necessary to control prosodic and spectral features, and so HMM speech synthesis [6] is ideally suited to meet this criteria. In this paper, an HMM synthesis system for various emotion styles is presented using two methodologies: (a) a mixed-style modelling where a full acoustic model is created containing all emotion styles; (b) an adaptation technique to convert a neutral style into a desired emotion using small amounts of data. These two techniques are compared through subjective and objective tests using three emotional speaking styles: neutral, happy and sad.

This paper is organized as follows: Section 2 introduces the HMM-TTS system, Section 3 summarizes the emotions database, Sections 4 and 5 describe the emotion modelling and adaption respectively and finally some concluding remarks are presented.

2   HMM-Based Speech Synthesis

The HMM-TTS used in this paper follows the structure originally proposed in [6] where vocal tract, F0 and state duration are simultaneously modelled. The standard system is based on the MLSA (Mel Log Spectrum Approximation) filter which uses a binary excitation signal (i.e. deltas plus noise). In order to increase the quality of the synthetic speech, it is necessary to reduce the buzzy effect of the vocoder and to apply a post-processing of the generated parameters to overcome the over-smoothing of the statistical and clustering processes [7]. Therefore, in order to alleviate these problems, a mixed excitation based on a multiband aperiodicity analysis is used (as described in the Blizzard system of Zen et al [8]) along with a post-filtering technique to improve the generated mel-cepstrum. The main features of the HMM-TTS system are described in the following sections.

2.1   Vocal Tract Modelling

The system uses a 39th-order mel-cepstral as the vocal tract representation computed from the spectrum of STRAIGHT [9] through a recursive algorithm [8]. The conventional mel-cepstral suffers from the F0 effect in the lower frequency range when the order is high because it captures the harmonics instead of the actual envelope. Using the F0 values, STRAIGHT carries out a F0-adaptive spectral analysis combined with a surface reconstruction method in the time-frequency region to remove signal periodicity.

2.2   Mixed Excitation

The excitation signal is designed as a weighted sum of a pulse train with phase manipulation and Gaussian noise. The weighting process is performed in the frequency domain using an aperiodicity measurement. This is based on a ratio between the lower and upper smoothed spectral envelopes and represent the relative energy distribution of aperiodic components [10]. In order to reliable statistically model this information is averaged on five frequency sub-bands (i.e., 0-1, 1-2, 2-4, 4-6, 6-8 KHz).

2.3   Contextual Factors

The baseline Spanish HMM-TTS system is described in detail in [11]. Phonetic and linguistic information is used as contextual factors to construct the clustering decision trees. Basically, the system uses information at different levels: syllable (length, position of phonemes), word (POS, position in utterance), utterance (accentual groups, related with the position of the stress in the words, and intonational groups to control interrogative, exclamative and declarative styles).

3   Evaluation of Speech Database

The emotional speech database used in this work was developed to learn the acoustic models of emotional speech and to be used as a database for the synthesizer. It was used in an objective evaluation conducted by means of automatic emotion identification techniques using statistical features obtained from the prosodic parameters of speech [12].

For the recording of the corpus, a female professional speaker was chosen as it was thought she could successfully convey the suitable expressive style to each text category (simulated/acted speech). For the design of texts semantically related to different expressive styles, an existing textual database of advertisements extracted from newspapers and magazines was used. Based on audio-visual publicity, three categories of the textual corpus were chosen and the most suitable emotion/style were assigned to them: New technologies (neutral), education (happy) and trips (sad-melancholic). It is important to mention that the speaker had previously received training in the vocal patterns (segmental and suprasegmental) of each style. The use of texts from an advertising category aims to help the speaker to maintain the desired style through the whole recording session. Therefore, the intended style was not performed according to the speaker's criteria for each sentence, but all the utterances of the same style were consecutively recorded in the same session following the previously learned pattern. The speaker was able to keep the required expressiveness even with texts whose semantic content were not coherent with the style.

For each style, 96 utterances were chosen and the test set was divided in several subsets. A forced answer test was designed with the question *What emotional*

Table 1. Average confusion matrix for the subjective test

Emotion	Happy	Sad	Neutral	Other
Happy	**81.0%**	0.1%	1.9%	17%
Sad	0.0%	**98.8%**	0.5%	0.7%
Neutral	2.3%	1.3%	**90.4%**	6%

state do you recognize from the voice of the speaker in this phrase?. The possible answers are the 3 styles of the corpus plus one more option for *other*. Table 1 shows the percentage of identification by style and test, being the sad style the best rated, followed by neutral and finally happy one.

4   Emotion Modelling

In order to model the emotional style, two methods were proposed in [5]. Firstly, style dependent modelling considers each style individually and a root node is used to gather all styles in one full model. The main advantage is that it is straightforward to add new a emotion. The second method used mixed style modelling, where emotions are clustered using a contextual factor together with the standard linguistic factors. In this case, all emotions are modelled in a single acoustic model though it is not easy to add a new emotion. However, since similar parameters are used among emotions the final model will be more accurate. Moreover it was shown in [5] that the naturalness of synthesized speech generated from style dependent and mixed models is almost the same, but the number of output distributions of the mixed style model is clearly smaller, so it is more attractive when an efficient solution is required. Because of this mixed style modelling is used in this work and a new contextual factor was added to the existing ones to distinguish the three emotional speaking styles.

5   Adaptation of Style Models

There are several possible adaptation techniques which use linear regression [13]. The physical manifestation of the various emotional expressive styles is complex and affects both vocal tract behaviour and prosody. The transformed parameters of the emotion adaptation must not be limited to only the mean vectors of the average voice model because the range of the variation is one of the important factors of the F0. In the standard Maximum Likelihood Linear Regression (MLLR) adaptation, mean vectors of state output distribution (b) for the target emotion style are obtained by linearly transforming mean vectors of state output distributions for the average style model. In order to get a better adaptation, a Constrained Maximum Likelihood Linear Regression (CMLLR) technique is used and mean and covariance matrices are obtained by transforming the parameters at the same time. In addition to the CMLLR adaptation, single bias removal and maximum a posteriori criterion are also used [14].

Considering $\mathcal{N}(\bullet\,;\bullet\,,\bullet\,)$ as the multivariate Gaussian distribution with mean $\bullet$ and covariance matrix $\bullet$ and $\bullet$ the vector of observations, we define the adapted output distribution for the i-th state as:

$$b_i(\bullet) = \mathcal{N}(\bullet\,;\bullet\bullet\,_i - \bullet\,,\bullet\bullet\,_i\bullet^\top) \tag{1}$$

where $\bullet = [\bullet,\bullet]$ is the transformation matrix with a linear transformation $\bullet$ and a bias vector $\bullet$.

6 Results

6.1 Subjective Evaluation

Three subjective evaluations over 20 test utterances were conducted to evaluate the performance of the systems (see Figures 1 and 2). The happy and the sad styles were used as target styles and adapted from the neutral style with 10, 30 and 200 utterances. The Full identifier stands for the mixed style modelling using a full corpus built with the happy, sad and neutral styles.

The first notable result is related to the naturalness of the synthesized speech measured in Figure 1. The first notable result shows the neutral emotion reaches a MOS (Mean Opinion Score) of 3.8 while the emotional styles vary higher and lower. The happy style is affected by high F0 values that distort the quality of the signal whereas the sad style has a better quality close to the natural speech with a MOS up to 4 for the style mixed modelling. It is a matter of fact that HMM-based synthesis over-smooths the generated parameters [15]. This effect is helping the sad style score since it has a lower variation of the F0.

Secondly, the two Figures (1 and 2) give a measure of how well the synthesizer actually reproduced each emotion style. The users were asked to score in a MOS test the intensity of the emotion being reproduced (i.e., happy and sad). We believe it is fair to conclude that any emotional style is properly reproduced when its score goes over 3. The results show that the happy style is more affected by the number of adaptation utterances than the sad style. In fact, the happy style is not perceived until the system is adapted with almost 30 utterances. In contrast, the sad style is perceived using just 10 utterances.

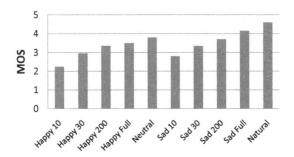

Fig. 1. MOS for naturalness

6.2 Objective Evaluation

An objective experiment is conducted to confirm the effect of the emotion style production[1]. In some works, a distortion of the mel-cepstrum is calculated to determine the acoustic distance between the neutral and the target emotional speech [16], whereas others present a root-mean-squares error of the F0 [13]. As

---

[1] Find examples here: http://www.salle.url.edu/~gonzalvo/hmm

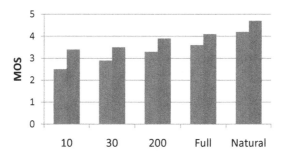

Fig. 2. MOS for emotion style intensity (sad in red and happy in blue)

the adaptation of emotional expressive styles affects both the vocal tract and the prosody, an evaluation to implicitly measure its performance would be desirable. VoQ parameters comply with this conditions and were shown to be useful for emotion discrimination [17]. These parameters are a set of measurements to weigh aesthetic features of the speech (e.g. harsh, trembling). The following VoQ parameters were directly computed over the vowels of 200 synthesized test utterances:

- Jitter (J) and Shimmer (S). Describe frequency and amplitude modulation noise. These parameters are calculated using the new model presented in [18] in order to minimize the effect of the prosody (see jitter in equation 2 and shimmer in equation 3 where i is the analyzed frame, $F0$ the fundamental frequency, N the size of the analysis window and ϕ the peak-to-peak amplitude variation between consecutive periods).

$$J_i = \frac{1}{N} \sum_{j=1}^{N-1} (F0_i(j+1) - F0_i(j))^2 \tag{2}$$

$$S_i = \frac{1}{N} \sum_{j=1}^{N-1} (\phi_i(j+1) - \phi_i(j))^2 \tag{3}$$

- Glottal-to-Noise Excitation Ratio (GNE) [19]. This parameter weighs the additive noise and it is a good alternative to Harmonic-to-Noise Ratio (HNR) parameter since it is almost independent from jitter and shimmer.
- Spectral Flatness Measure (SFM), computed as the ratio of the geometric to the arithmetic mean of the spectral energy distribution. In equation 4, E_i is the energy in the frame i.

$$SFM_i = 10 \log \frac{\sqrt[N]{\prod_{i=1}^{N} E_i}}{\frac{\prod_{i=1}^{N} E_i}{N}} \tag{4}$$

- Hammarberg Index (HammI), defined as the difference between the maximum energy in the 0-2000 Hz and 2000-5000 Hz frequency bands (see equation 5).

$$HammI = 10 \log \frac{\max(E_{0-2000Hz})}{\max(E_{2000Hz-5000Hz})} \tag{5}$$

- Drop-off of spectral energy above 1000Hz (Drop_1000), a linear approximation of spectral tilt above 1000 Hz, which is calculated using a least squares method [20].
- Relative amount of energy in the high (above 1000Hz) versus the low frequency range of the voice spectrum (see Pe_1000 in equation 6 where f_s is the sampling rate and E_f is the energy in the frequency band).

$$Pe_{1000} = 10 \log \frac{\sum_{f=1000Hz}^{F_s/2} E_f}{\sum_{f=0}^{1000Hz} E_f} \tag{6}$$

It is not the aim of this paper to discriminate the synthesized emotions but to show the limitations of the synthetic styles. Thus Figures 3 and 4 shows a descriptive statistics of the VoQ parameters for the happy and sad styles (each column represents a parameter). In this figure, the area within lines represents the standard deviation of the natural speech. It can be seen that most of the VoQ parameters are inside the standard deviation. Nevertheless, note that the HammI parameters for the sad style are biased towards high values which means that the synthetic is more low-pitched than the natural one. Also, it is important to highlight the GNE parameter for the happy style. This indicates a sort of distortion that can be considered as an additive noise (note that the subjective measure coincide in that the happy style is affected by a distortion in such a manner its MOS for naturalness is below the other styles). Finally, the jitter results for the sad style indicates that the synthesized speech suffers from a trembling effect (concretely, the more utterances to adapt, the higher the jitter).

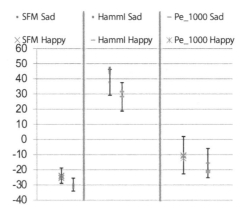

Fig. 3. Main VoQ parameters for happy style

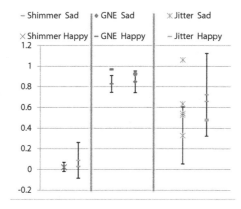

Fig. 4. Main VoQ parameters for sad style

7   Conclusions

In this work we have presented two techniques for the synthesis of Spanish emotion speaking styles using a HMM-TTS system. First, we presented the features of the system being used (mixed excitation based on aperiodicity and mel-cepstrum extracted from the STRAIGHT spectrum). Secondly, we described the database and its evaluation. Then we investigated the use of two known techniques, one for modelling the emotion in an HMM system and another for adaptation to create emotional styles synthesis with a small amounts of data. From results of subjective experiments we have shown the quality of the synthetic speech and we have evaluated the intensity of the target emotion style. From objective measures we have shown the effectiveness of the proposed approaches. The future plan is to increase the number of emotion styles and use a more accurate independent style voice.

References

1. Bulut, M., Narayanan, S., Syrdal, A.: Expressive speech synthesis using a concatenative synthesizer. In: Proc. of ICSLP, Denver, USA (September 2002)
2. Montero, J.M., Guiterrez-Arriola, J., Colas, J., Macias, J., Enriquez, E., Pardo, J.M.: Development of an emotional speech synthesizer in spanish. In: Proc of Eurospeech, Budapest, pp. 2099–2102 (1999)
3. Inanoglu, Z., Young, S.: A system for transforming the emotion in speech: Combining data-driven conversion techniques for prosody and voice quality. In: Proc. of Interspeech, Antwerp, Belgium (August 2007)
4. Tamura, M., Masuko, T., Tokuda, K., Kobayashi, T.: Text-to-speech synthesis with arbitrary speaker's voice from average voice from average voice. In: Proc. of Eurospeech (2001)
5. Yamagishi, J., Masuko, T., Kobayashi, T.: HMM-based expressive speech synthesis – towards tts with arbitrary speaking styles and emotions. In: Proc. of Special Workshop in Maui, SWIM (2004)

6. Yoshimura, T., Tokuda, K., Masuko, T., Kobayashi, T., Kitamura, T.: Simultaneous modeling of spectrum, pitch and duration in HMM-based speech synthesis. In: Proc. of Eurospeech (1999)
7. Toda, T., Tokuda, K.: A speech parameter generation algorithm considering global variance for HMM-based speech synthesis. IEICE Transactions on Fundamentals of Electronics, Communications and Computer Sciences E90-D 5, 816–824 (2007)
8. Zen, H., Toda, T., Nakamura, M., Tokuda, K.: Details of nitech HMM-based speech synthesis system for the blizzard challenge 2005. IEICE Transactions on Fund. of Electronics, Comm. and Computer Sciences E90-D 1, 325–333 (2007)
9. Kawahara, H., Estill, J., Fujimura, O.: Aperiodicity extraction and control using mixed mode excitation and group delay manipulation for a high quality speech analysis, modification and synthesis system straight. In: Proc. of MAVEBA (2001)
10. Kawahara, H.: Speech representation and transformation using adaptive interpolation of weighted spectrum: Vocoder revisited. In: Proc. of ICASSP, Washington, DC, USA, p. 1303. IEEE Computer Society, Los Alamitos (1997)
11. Gonzalvo, X., Socoró, J., Iriondo, I., Monzo, C., Martínez, E.: Linguistic and mixed excitation improvements on a HMM-based speech synthesis for castilian spanish. In: Proc. of ICASSP, Bonn, Germany (2007)
12. Iriondo, I., Planet, S., Socoro, J., Alias, F.: Objective and subjective evaluation of an expressive speech corpus. In: Proc. of NoLISP, Paris, France (May 2007)
13. Yamagishi, J., Ogata, K., Nakano, Y., Isogai, J., Kobayashi, T.: HSMM-based model adaptation algorithms for average-voice-based speech synthesis. In: Proc. of ICASSP, Toulouse, France (2006)
14. Yamagishi, J., Kobayashi, T.: Hidden Semi-Markov model and its speaker adaptation techniques. IEICE Transactions on Audio, Speech and Language Processing 6 (2007)
15. Toda, T., Tokuda, K.: Speech parameter generation algorithm considering global variance for HMM-based speech synthesis. In: Proc. of Interspeech, Portugal, pp. 2801–2804 (2005)
16. Kawanami, H., Iwami, Y., Toda, T., Saruwatari, H., Shikano, K.: GMM-based voice conversion applied to emotional speech synthesis. In: Proc. of Eurospeech, Geneva, Switzerland (September 2003)
17. Monzo, C., Alias, F., Iriondo, I., Gonzalvo, X., Planet, S.: Discriminating expressive speech styles by voice quality parameterization. In: Proc. of ICPhS (2007)
18. Monzo, C., Iriondo, I., Martínez, E.: Procedimiento para la medida y la modificación del jitter y del shimmer aplicado a la síntesis del habla expresiva. In: Proc. of JTH, Bilbao, Spain (2008)
19. Michaelis, D., Gramss, T., Strube, H.W.: Glottal to noise excitation ratio - a new measure for describing pathological voices. In: Acustica, Acta Acustica, pp. 700–706 (1997)
20. Abdi, H.: Least Squares. In: Lewis-Beck, M., Bryman, A., Futing, T. (eds.) Encyclopedia for research methods for the social sciences, pp. 559–561. Sage, Thousand Oaks (2003)

Glottal Source Estimation Using an Automatic Chirp Decomposition

Thomas Drugman[1], Baris Bozkurt[2], and Thierry Dutoit[1]

[1] TCTS Lab, Faculté Polytechnique de Mons, Belgium
[2] Department of Electrical & Electronics Engineering, Izmir Institute of Technology, Turkey

Abstract. In a previous work, we showed that the glottal source can be estimated from speech signals by computing the Zeros of the Z-Transform (ZZT). Decomposition was achieved by separating the roots inside (causal contribution) and outside (anticausal contribution) the unit circle. In order to guarantee a correct deconvolution, time alignment on the Glottal Closure Instants (GCIs) was shown to be essential. This paper extends the formalism of ZZT by evaluating the Z-transform on a contour possibly different from the unit circle. A method is proposed for determining automatically this contour by inspecting the root distribution. The derived Zeros of the Chirp Z-Transform (ZCZT)-based technique turns out to be much more robust to GCI location errors.

1 Introduction

The deconvolution of speech into its vocal tract and glottis contributions is an important topic in speech processing. Explicitly isolating both components allows to model them independently. While techniques for modeling the vocal tract are rather well-established, it is not the case for the glottal source representation. However the characterization of this latter has been shown to be advantageous in speaker recognition [1], speech disorder analysis [2], speech recognition [3] or speech synthesis [4]. These reasons justify the need of developing algorithms able to robustly and reliably estimate and parametrize the glottal signal.

Some works addressed the estimation of the glottal contribution directly from speech waveforms. Most approaches rely on a first parametric modeling of the vocal tract and then remove it by inverse filtering so as to obtain the glottal signal estimation. In [5], the use of the Discrete All-Pole (DAP) model is proposed. The Iterative Adaptive Inverse Filtering technique (IAIF) described in [6] isolates the source signal by iteratively estimating both vocal tract and source parts. In [7], the vocal tract is estimated by Linear Prediction (LP) analysis on the closed phase. As an extension, the Multicycle closed-phase LPC (MCLPC) method [8] refines its estimation on several larynx cycles. In a fundamentally different point of view, we proposed in [9] a non-parametric technique based on the Zeros of the Z-Transform (ZZT). ZZT basis relies on the observation that speech is a mixed-phase signal [10] where the anticausal component corresponds to the vocal folds open phase, and where the causal component comprises both the glottis

J. Solé-Casals and V. Zaiats (Eds.): NOLISP 2009, LNAI 5933, pp. 35–42, 2010.

closure and the vocal tract contributions. Basically ZZT isolates the glottal open phase contribution from the speech signal, by separating its causal and anticausal components. In [11], a comparative evaluation between LPC and ZZT-based decompositions is led, giving a significant advantage for the second technique.

This paper proposes an extension to the traditional ZZT-based decomposition technique. The new method aims at separating both causal and anticausal contributions by computing the Zeros of a Chirp Z-Transform (ZCZT). More precisely, the Z-transform is here evaluated on a contour possibly different from the unit circle. As a result, we will see that the estimation is much less sensitive to the Glottal Closure Instant (GCI) detection errors. In addition, a way to automatically determine an optimal contour is also proposed.

The paper is structured as follows. Section 2 reminds the principle of the ZZT-based decomposition of speech. Its extension making use of a chirp analysis is proposed and discussed in Section 3. In Section 4, a comparative evaluation of both approaches is led on both synthetic and real speech signals. Finally we conclude in Section 5.

2   ZZT-Based Decomposition of Speech

For a series of N samples $(x(0), x(1), ...x(N-1))$ taken from a discrete signal $x(n)$, the ZZT representation is defined as the set of roots (zeros) $(Z_1, Z_2, ...Z_{N-1})$ of the corresponding Z-Transform $X(z)$:

$$X(z) = \sum_{n=0}^{N-1} x(n)z^{-n} = x(0)z^{-N+1} \prod_{m=1}^{N-1} (z - Z_m) \tag{1}$$

The spectrum of the glottal source open phase is then computed from zeros outside the unit circle (anticausal component) while zeros inside it give the vocal tract transmittance modulated by the source return phase spectrum (causal component). To obtain such a separation, the effects of the windowing are known to play a crucial role [12]. In particular, we have shown that a Blackman window centered on the Glottal Closure Instant (GCI) and whose length is twice the pitch period is appropriate in order to achieve a good decomposition.

3   Chirp Decomposition of Speech

The Chirp Z-Transform (CZT), as introduced by Rabiner et al [13] in 1969, allows the evaluation of the Z-transform on a spiral contour in the Z-plane. Its first application aimed at separating too close formants by reducing their bandwidth. Nowadays CZT reaches several fields of Signal Processing such as time interpolation, homomorphic filtering, pole enhancement, narrow-band analysis,...

As previously mentioned, the ZZT-based decomposition is strongly dependent on the applied windowing. This sensitivity may be explained by the fact that ZZT implicitly conveys phase information, for which time alignment is known to be crucial [14]. In that article, it is observed that the window shape and onset may lead to zeros whose topology can be detrimental for accurate pulse estimation. The subject of this work is precisely to handle with these zeros close to the unit circle, such that the ZZT-based technique correctly separates the causal (i.e minimum-phase) and anticausal (i.e maximum-phase) components.

For this, we evaluate the CZT on a circle whose radius R is chosen so as to split the root distribution into two well-separated groups. More precisely, it is observed that the significant impulse present in the excitation at the GCI results in a gap in the root distribution. When analysis is exactly GCI-synchronous, the unit circle perfectly separates causal and anticausal roots. On the opposite, when the window moves off from the GCI, the root distribution is transformed. Such a decomposition is then not guaranteed for the unit circle and another boundary is generally required. Figure 1 gives an example of root distribution for a natural voiced speech frame for which an error of 0.6 ms is made on the real GCI position. It is clearly seen that using the traditional ZZT-based decomposition ($R = 1$) for this frame will lead to erroneous results. In contrast, it is possible to find an optimal radius leading to a correct separation.

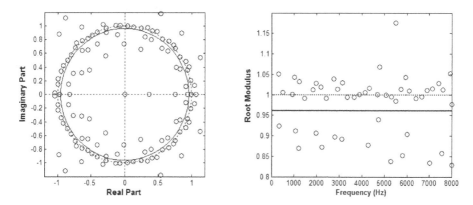

Fig. 1. Example of root distribution for a natural speech frame. *Left panel*: representation in the Z-plane, *Right panel*: representation in polar coordinates. The chirp circle (solid line) allows a correct decomposition, contrarily to unit circle (dotted line).

In order to automatically determine such a radius, let us have the following thought process. We know that ideally the analysis should be GCI-synchronous. When this is not the case, the chirp analysis tends to modify the window such that its center coincides with the nearest GCI (to ensure a reliable phase information). Indeed, evaluating the chirp Z-transform of a signal $x(t)$ on a circle of radius R is equivalent to evaluating the Z-transform of $x(t) \cdot exp(log(1/R) \cdot t)$ on the unit circle. It can be demonstrated that for a Blackman window $w(t)$ of length L starting in $t = 0$:

$$w(t) = 0.42 - 0.5 \cdot \cos(\frac{2\pi t}{L}) + 0.08 \cdot \cos(\frac{4\pi t}{L}), \qquad (2)$$

the radius R necessary to modify its shape so that its new maximum lies in position t^* ($< L$) is expressed as:

$$R = exp[\frac{2\pi}{L} \cdot \frac{41\tan^2(\frac{\pi t^*}{L}) + 9}{25\tan^3(\frac{\pi t^*}{L}) + 9\tan(\frac{\pi t^*}{L}))}]. \tag{3}$$

In particular, we verify that $R = 1$ is optimal when the window is GCI-centered ($t^* = \frac{L}{2}$) and, since we are working with two-period long windows, the optimal radius does not exceed $exp(\pm\frac{50\pi}{17L})$ in the worst cases (the nearest GCI is then positioned in $t^* = \frac{L}{4}$ or $t^* = \frac{3L}{4}$). As a means for automatically determining the radius allowing an efficient separation, the sorted root moduli are inspected and the greatest discontinuity in the interval $[exp(-\frac{50\pi}{17L}), exp(\frac{50\pi}{17L})]$ is detected. Radius R is then chosen as the middle of this discontinuity, and is assumed to optimally split the roots into minimum and maximum-phase contributions.

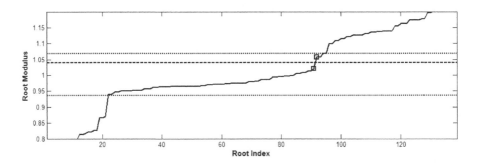

Fig. 2. Determination of radius R (dashed line) for ZCZT computation by detecting, within the bounds $exp(\pm\frac{50\pi}{17L})$ (dotted lines), a discontinuity (indicated by rectangles) in the sorted root moduli (solid line)

4   Experimental Results

This Section gives a comparative evaluation of the following methods:

- *the traditional ZZT-based technique*: $R = 1$,
- *the proposed ZCZT-based technique*: R is computed as explained at the end of Section 3 (see Fig. 2),
- *the ideal ZCZT-based technique*: R is computed from Equation 3 where the real GCI location t^* is known. This can be seen as the ultimate performance one can expect from the ZCZT-based technique.

Among others, it is emphasized how the proposed technique is advantageous in case of GCI location errors.

4.1   Results on Synthetic Speech

Objectively and quantitatively assessing a method of glottal signal estimation requires working with synthetic signals, since the real source is not available for

Table 1. Details of the test conditions for the experiments on synthetic signals

Source Characteristics			Filter	Perturbation
Open Quotient	Asymmetry Coeff.	Pitch	Vowel	GCI location error
0.4:0.05:0.9	0.6:0.05:0.9	60:20:180 Hz	/a/,/e/,/i/,/u/	-50:5:50 % of T_0

real speech signals. In this work, synthetic speech signals are generated for different test conditions, by passing a train of Liljencrants-Fant waves [15] through an all-pole filter. This latter is obtained by LPC analysis on real sustained vowel uttered by a male speaker. In order to cover as much as possible the diversity one can find in real speech, parameters are varied over their whole range. Table 1 summarizes the experimental setup. Note that since the mean pitch during the utterances for which the LP coefficients were extracted was about 100 Hz, it reasonable to consider that the fundamental frequency should not exceed 60 and 180 Hz in continuous speech.

To evaluate the performance of our methods, two objective measures are used:

– the determination rate on the glottal formant frequency F_g: As one of the main feature of the glottal source, the glottal formant [10] should be preserved after estimation. The determination rate consists of the percentage of frames for which the relative error made on F_g is lower than 20%.
– the spectral distortion: This measure quantifies in the frequency-domain the distance between the reference and estimated glottal waves (here noted x and y by simplification), expressed as:

$$SD(x,y) = \sqrt{\int_{-\pi}^{\pi} (20\log_{10}|\frac{X(\omega)}{Y(\omega)}|)^2 \frac{d\omega}{2\pi}} \qquad (4)$$

Figure 3 compares the results obtained for the three methods according to their sensitivity to the GCI location. The proposed ZCZT-based technique is clearly seen as an enhancement of the traditional ZZT approach when an error on the exact GCI position is made.

4.2   Results on Real Speech

Figure 4 displays an example of decomposition on a real voiced speech segment (vowel /e/ from *BrianLou4.wav* of the Voqual03 database, $F_s = 16kHz$). The top panel exhibits the speech waveform together with the synchronized (compensation of the delay between the laryngograph and the microphone) differenced Electroglottograph (EGG) informative about the GCI positions. Both next panels compare respectively the detected glottal formant frequency F_g and the radius for the three techniques. In the middle panel, deviations from the constant F_g can be considered as errors since F_g is expected to be almost constant during three pitch periods. It may be noticed that the traditional ZZT-based method degrades if analysis is not achieved in the GCI close vicinity. Contrarily, the

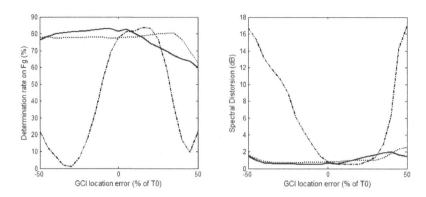

Fig. 3. Comparison of the traditional ZZT (dashdotted line), proposed ZCZT (solid line) and ideal ZCZT (dotted line) based methods on synthetic signals according to their sensitivity to an error on the GCI location. *Left panel:* Influence on the determination rate on the glottal formant frequency. *Right panel:* Influence on the spectral distortion.

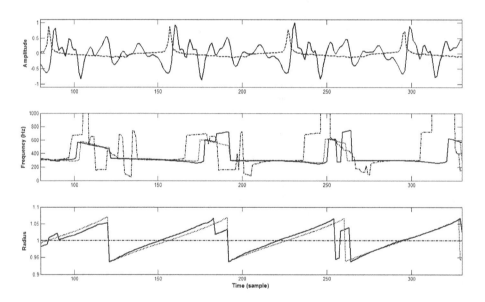

Fig. 4. Comparison of ZZT and ZCZT-based methods on a real voiced speech segment. *Top panel:* the speech signal (solid line) with the synchronized differenced EGG (dashed line). *Middle panel:* the glottal formant frequency estimated by the traditional ZZT (dashdotted line), the proposed ZCZT (solid line) and the ideal ZCZT (dotted line) based techniques. *Bottom panel:* Their corresponding radius used to compute the chirp Z-transform.

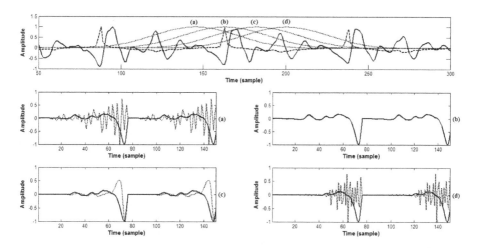

Fig. 5. Examples of glottal source estimation using either the traditional ZZT or the proposed ZCZT-based method. *Top panel:* a voiced speech segment (solid line) with the synchronized differenced EGG (dashed line) and four different positions of the window (dotted line). *Panels (a) to (d):* for the corresponding window location, two cycles of the glottal source estimation achieved by the traditional ZZT (dotted line) and by the proposed ZCZT-based technique (solid line).

proposed ZCZT-based technique gives a reliable estimation of the glottal source on a large segment around the GCI. Besides the obtained performance is comparable to what is carried out by the ideal ZCZT. In Figure 5 the glottal source estimated by the traditional ZZT and the proposed ZCZT-based method are displayed for four different positions of the window (for the vowel /a/ from the same file). It can be observed that the proposed technique (solid line) gives a reliable estimation of the glottal flow wherever the window is located. On the contrary the sensivity of the traditional approach can be clearly noticed since its glottal source estimation turns out to be irrelevant when the analysis is not performed in a GCI-synchronous way.

5 Conclusion

This paper proposed an extension of the ZZT-based technique we proposed in [9]. The enhancement consists in evaluating the Z-transform on a contour possibly different from the unit circle. For this we considered, in the Z-plane, circles whose radius is automatically determined by detecting a discontinuity in the root distribution. It is expected that such circles lead to a better separation of both causal and anticausal contributions. Results obtained on synthetic and real speech signals report an advantage for the proposed ZCZT-based technique, mainly when GCIs are not accurately localized. As future work, we plan to characterize the glottal source based on the proposed framework.

Acknowledgments

Thomas Drugman is supported by the "Fonds National de la Recherche Scientifique" (FNRS).

References

1. Plumpe, M., Quatieri, T., Reynolds, D.: Modeling of the glottal flow derivative waveform with application to speaker identification. IEEE Trans. on Speech and Audio Processing 7, 569–586 (1999)
2. Moore, E., Clements, M., Peifer, J., Weisser, L.: Investigating the role of glottal features in classifying clinical depression. In: Proc. of the 25th International Conference of the IEEE Engineering in Medicine and Biology Society, vol. 3, pp. 2849–2852 (2003)
3. Yamada, D., Kitaoka, N., Nakagawa, S.: Speech Recognition Using Features Based on Glottal Sound Source. Trans. of the Institute of Electrical Engineers of Japan 122-C(12), 2028–2034 (2002)
4. Drugman, T., Wilfart, G., Moinet, A., Dutoit, T.: Using a pitch-synchronous residual for hybrid HMM/frame selection speech synthesis. In: Proc. IEEE International Conference on Speech and Signal Processing (2009)
5. Alku, P., Vilkman, E.: Estimation of the glottal pulseform based on discrete all-pole modeling. In: Third International Conference on Spoken Language Processing, pp. 1619–1622 (1994)
6. Alku, P., Svec, J., Vilkman, E., Sram, F.: Glottal wave analysis with pitch synchronous iterative adaptive inverse filtering. Speech Communication 11(2-3), 109–118 (1992)
7. Veeneman, D., BeMent, S.: Automatic glottal inverse filtering from speech and electroglottographic signals. IEEE Trans. on Signal Processing 33, 369–377 (1985)
8. Brookes, D., Chan, D.: Speaker characteristics from a glottal airflow model using glottal inverse filtering. Proc. Institue of Acoust. 15, 501–508 (1994)
9. Bozkurt, B., Doval, B., D'Alessandro, C., Dutoit, T.: Zeros of Z-Transform Representation With Application to Source-Filter Separation in Speech. IEEE Signal Processing Letters 12(4) (2005)
10. Doval, B., d'Alessandro, C., Henrich, N.: The voice source as a causal/anticausal linear filter. In: Proceedings ISCA ITRW VOQUAL 2003, pp. 15–19 (2003)
11. Sturmel, N., D'Alessandro, C., Doval, B.: A comparative evaluation of the Zeros of Z-transform representation for voice source estimation. In: The Interspeech 2007, pp. 558–561 (2007)
12. Bozkurt, B., Doval, B., D'Alessandro, C., Dutoit, T.: Appropriate windowing for group delay analysis and roots of Z-transform of speech signals. In: Proc. of the 12th European Signal Processing Conference (2004)
13. Rabiner, L., Schafer, R., Rader, C.: The Chirp-Z transform Algorithm and Its Application. Bell System Technical Journal 48(5), 1249–1292 (1969)
14. Tribolet, J., Quatieri, T., Oppenheim, A.: Short-time homomorphic analysis. In: IEEE International Conference on Speech and Signal Processing, vol. 2, pp. 716–722 (1977)
15. Fant, G., Liljencrants, J., Lin, Q.: A four parameter model of glottal flow. STL-QPSR4, 1–13 (1985)

Automatic Classification of Regular vs. Irregular Phonation Types

Tamás Bőhm, Zoltán Both, and Géza Németh

Department of Telecommunications and Media Informatics,
Budapest University of Technology and Economics,
Magyar Tudósok krt. 2., 1117 Budapest, Hungary
{bohm,bothzoli,nemeth}@tmit.bme.hu

Abstract. Irregular phonation (also called creaky voice, glottalization and laryngealization) may have various communicative functions in speech. Thus the automatic classification of phonation type into regular and irregular can have a number of applications in speech technology. In this paper, we propose such a classifier that extracts six acoustic cues from vowels and then labels them as regular or irregular by means of a support vector machine. We integrated cues from earlier phonation type classifiers and improved their performance in five out of the six cases. The classifier with the improved cue set produced a 98.85% hit rate and a 3.47% false alarm rate on a subset of the TIMIT corpus.

Keywords: Irregular phonation, creaky voice, glottalization, laryngealization, phonation type, voice quality, support vector machine.

1 Introduction

Voiced speech can be characterized by the regular vibration of the vocal folds, resulting in a quasi-periodic speech waveform (i.e. the length, the amplitude and the shape of adjacent periods show only slight differences). However, sometimes the vocal folds vibrate irregularly and thus successive cycles exhibit abrupt, substantial changes in their length, amplitude or shape (Fig. 1). In this paper, the term irregular phonation is used to describe regions of speech that display "either an unusual difference in time or amplitude over adjacent pitch periods that exceeds the small-scale jitter and shimmer differences, or an unusually wide spacing of the glottal pulses compared to their spacing in the local environment" [1]. This latter case refers to periodic vibration well below the speaker's normal fundamental frequency range. Phonation types, in which the irregularity arises from additive noise (e.g. breathiness) are not considered in this study.

It is likely that irregular phonation plays a role in various aspects of speech communication. It has been shown to be a cue to segmental contrasts and to prosodic structure in several languages [2]. Further, its occurrence seems to be characteristic to certain speakers [3] and to certain emotional states [4]. An automatic method for classifying phonation into regular and irregular can help the analysis of these communicative roles and can also be useful in improving speech technologies (e.g. speech recognition, emotional state classification and speaker identification).

J. Solé-Casals and V. Zaiats (Eds.): NOLISP 2009, LNAI 5933, pp. 43–50, 2010.

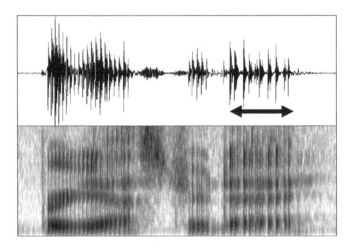

Fig. 1. Waveform (top) and spectrogram (bottom) of a speech signal exhibiting irregular phonation (denoted by the arrow)

A number of phonation type classifiers employing a wide spectrum of acoustic cues and decision algorithms are described in the literature. The method of Surana and Slifka [1,5] calculates four acoustic cues (e.g. fundamental frequency and normalized amplitude) and then applies a support vector machine (SVM) classification scheme to decide whether a phone was produced with regular or irregular phonation. Ishi et al. [6] proposed three cues based on the peaks of the very-short-term power of the speech signal (e.g. cues related to the rate of power change around the peaks and the periodicity between the peaks). The final decision is based on thresholds for the values of the three cues. Vishnubhotla and Espy-Wilson [7] employed cues derived from the AMDF (average magnitude difference function) dip profile, as well as zero-crossing rate, spectral slope and pitch detection confidence (autocorrelation peak value). Yoon et al. [8] also used the peak value of the autocorrelation function and measured open quotient, too (by means of the amplitude difference between the first and second harmonics). Kiessling et al. [9] used five acoustic cues extracted from the cepstrally smoothed spectrum and classified phonation by means of a Gaussian phone component recognizer. The other approach presented in the same paper employs an artificial neural network (ANN) to inverse filter the speech signal, and then another ANN to classify the source signals into regular, irregular and unvoiced.

In this paper, we propose a phonation type classifier that can categorize vowels either as regular or as irregular with high accuracy. Our method integrates cues from two earlier systems: all the four cues from [1] (fundamental frequency, normalized RMS amplitude, smoothed-energy-difference amplitude and shift-difference amplitude) and two cues from [6] (power peak rising and falling degrees, and intraframe periodicity). These six cues were reimplemented based on their description in the literature and most of them were improved by algorithmic refinements and by exploring their parameter space. The performance of each cue was measured by the increase in the area under the receiver operating characteristic (ROC) curve [10]. These cues are inputted into a support vector machine in order to classify the vowel as regular or irregular.

2 Speech Data Set

This work was carried out using the subset of the TIMIT corpus in which occurrences of irregular phonation in vowels were hand-labelled by Surana and Slifka [1,5]. This subset included recordings of 151 speakers (both males and females) uttering 10 sentences each (114 speakers in the train set and another 37 speakers in the test set). Among the vowels labeled, they found 1751 produced with irregular phonation and 10876 with regular phonation.

3 Acoustic Cues

For the extraction of five of the acoustic cues, the input speech signal is spliced into 30 ms frames with a 5 ms step. These five cues are expressed as a summary statistic (e.g. mean or minimum) over the values calculated for each frame in the input. The remaining one cue (power peaks) processes the entire input vowel in one run. For each cue, we first present its original calculation method (as described in the corresponding paper) and then we explain our improvements (if there was any).

3.1 Fundamental Frequency (F0)

One can expect that for irregularly phonated speech, a pitch detector extracts an F_0 value that is lower than that for regular voiced speech, or detects it as unvoiced (denoted by $F_0=0$ Hz). Surana and Slifka [1,5] used an autocorrelation-based pitch detector to calculate this acoustic cue. Before computing the autocorrelation function, the speech signal is Hamming-windowed and inverse filtered (by means of a 12th order LPC filter) and the residual signal is low-pass filtered with a 1 kHz cutoff. The peaks of the normalized autocorrelation function of the low-passed residual are used to estimate F_0:

- If there are no autocorrelation peaks higher than the voicing threshold (0.46) in the lag interval corresponding to 70-400 Hz, then the frame is considered unvoiced and F_0 is set to 0.
- If there is only one peak fulfilling the above criterion, then the reciprocal of the peak's lag is returned as the F_0.
- If there are more than one such peaks and their lags are integer multiples of each other, then the second one is chosen as the one corresponding to F_0 ("second peak rule").
- If there is no such regularity among the multiple peaks, then the highest peak is chosen.

For a given vowel, the F0 cue value is the minimum of the F_0's of the frames comprising that vowel.

By some changes in the algorithm, we could increase the area under the ROC curve of the cue from 0.87 to 0.93 (Fig. 2.a). These changes included the removal of the second peak rule as it usually resulted in halving errors, the use of an unbiased autocorrelation function (whose envelope does not decrease to zero at large lags [11]).

Further, the voicing threshold was changed to 0.35 based on a systematic evaluation of a number of values in the range of 0.25-0.5.

3.2 Normalized RMS Amplitude (NRMS)

In irregularly phonated speech, there are generally fewer glottal pulses in a given time frame than in regularly phonated speech. This can be captured by the RMS amplitude (intensity), normalized by the RMS amplitude of a longer interval. Surana and Slifka [1] divided the RMS intensity of each frame by the intensity calculated over the entire sentence and then took the mean of these values in order to compute the NRMS cue for the vowel.

If the intensity level changes along the sentence, then normalizing with the sentence RMS can become misleading. Thus instead of the entire sentence, we used the vowel and its local environment for normalization. In the range examined (25-500 ms, with 25 ms steps), an environment of 50 ms (25 ms before and 25 ms after) led to a small increase in the area under the ROC curve (0.84 to 0.86; Fig. 2.b).

3.3 Smoothed-Energy Difference Amplitude (SED)

This cue attempts to characterize the rapid energy transitions in irregular phonation that are due to the wider spacing of the glottal pulses. The energy curve (calculated as the mean magnitude between 300 and 1500 Hz of the FFT in 16 ms windows, stepped by 1 ms) is smoothed with a 16 ms and separately with a 6 ms window. The difference of the two smoothed energy functions is usually near zero for regular phonation, while in case of irregular phonation, it has several peaks. In the original description, the SED cue is the maximum of the difference function [1]. (The first and last 8 ms of the difference function is not taken into account because artifacts due to the different window sizes can appear in these regions.)

Instead of taking the maximum, we calculate the absolute maximum. Further, our tests examining the effect of various window size combinations revealed that using a 2 ms and a 4 ms smoothing window can improve the separation of regular and irregular tokens. After these changes, the area under the ROC curve increased by 0.08 (from 0.74 to 0.82; Fig. 2.c).

3.4 Shift-Difference Amplitude (SDA)

This cue estimates the irregularity of the glottal pulses. After preprocessing, two 10 ms windows are employed: one of them is initially shifted 2 ms to the left, while the other one is shifted 2 ms to the right of the center of the 30 ms frame [1]. In each step, both of the windows are shifted one sample away from the frame center and the squared difference signal of the two windowed waveforms is calculated. Then, the minimum of the difference signals is computed in each time point over all the window shifts. The resulting 10 ms signal is normalized by the square of the middle 10 ms of the frame and then averaged over time. The SDA cue for the vowel is the average of the SDA values obtained for each frame.

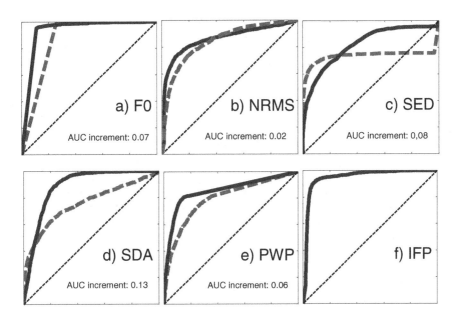

Fig. 2. Receiver operating characteristic (ROC) curves of the six acoustic cues: a) fundamental frequency, b) normalized RMS amplitude, c) smoothed-energy difference amplitude, d) shift-difference amplitude, e) power peak rising and falling degree, f) intraframe periodicity. The red dashed lines correspond to the original versions of the cues, while the blue continuous lines represent the improved versions (there were no improvements for the IFP cue). Both axes range from 0 to 1 on all panes. The increment in the area under the curves is also shown.

The shift-difference amplitude cue in [1] is based on Kochanski's "aperiodicity measure" [12]. According to our tests, applying the aperiodicity measure (with slight changes) results in higher separation between regular and irregular vowels than the SDA cue in [1] (the area under the ROC curve is 0.88, compared to 0.75 with the method described in [1]; Fig 2.d).

To calculate the "aperiodicity measure", we add 5 ms of silence to the beginning and to the end of the 30 ms frame. Then, as described in [12], we define two 30 ms windows: one at the beginning and the other one at the end of the zero-padded signal. As we shift both windows sample-by-sample towards the center, we calculate their squared difference function in each step. After down-sampling each difference function to 125 Hz (so that there are only 5 values remaining for each 40 ms long signal), the minimum value is taken at each time point. After dropping the first and the last value, these minimums are averaged to obtain the aperiodicity measure for the frame.

3.5 Power Peak Rising and Falling Degrees (PWP)

Like the smoothed-energy difference amplitude, this cue also attempts to capture the rapid energy transitions in irregular phonation. As described by Ishi et al. [6], the "very-short-term power contour" is calculated, with a window size of 4 ms and a step size of 2 ms (note that this cue is not calculated using the 30 ms framing employed by all the other cues). In this function, irregular glottal pulses usually appear as peaks with a long, steep rise and fall. Thus first the peaks are detected and then a measure of the rate of power rise and fall (before and after the peak) is computed. A point in the very-short-term power contour is considered to be a power peak, if it has a value of at least 2 dB higher than both the point 3 samples before and the point 3 samples later. The degree of power rising before the peak is obtained as the maximum power difference between the peak and the 5 preceding values. The power falling was estimated similarly, with the 5 following values. When adapted to our phonation type classification framework, the PWP cue is calculated as the average of the maximum power rising and the maximum power falling value in the vowel.

According to a grid search we performed, it is more advantageous to find the peaks based on the 4th value (instead of the 3rd) before and after the point in question. Further, the power rising and falling features separate the two phonation types better, if they are computed as the maximum power difference in a ±4 sample environment of the peak. The ROC curve of the original PWP cue has an area of 0.79, while the area under the curve corresponding to the improved version is 0.85 (Fig 2.e).

3.6 Intraframe Periodicity (IFP)

As the shift-difference amplitude, IFP also tries to capture the repeatable waveform structure of regular phonation and the reduction in this repeatability in irregular phonation. For each frame, the unbiased autocorrelation function is calculated and the first prominent peak with a positive lag is found [6]. Then the autocorrelation values at integer multiples of this peak lag are obtained and their minimum is selected as the measure of intraframe periodicity. For white noise, this value is near zero, while for a perfectly periodic signal, it is one. The IFP cue of the vowel is the maximum of IFP values of the individual frames. This cue provides excellent separation between regular and irregular vowels (the area under ROC curve is 0.95; Fig 2.f) and thus we did not attempt to improve it.

4   Classification

Using the six described cues (their improved versions, where available) as features, the classification was carried out by a support vector machine (SVM) with a radial basis function (RBF) kernel [13]. The SVM was implemented using the publicly available OSU SVM toolbox (http://sourceforge.net/projects/svm).

Before training, two parameters of the SVM needed to be set: C, the cost of incorrect classifications, and γ, a property of the Gaussian kernel. We ran a grid search on a wide range of values for the two parameters. A subset of the training set, containing 1380 regular and 1380 irregular vowels, was used for the grid search. For each parameter combination, we performed both a 3-fold and a 10-fold cross-validation (e.g. for the 3-fold cross-validation, we trained the SVM three times, always leaving out one third of the set that we subsequently used for testing the performance of the classifier) and calculated the average of the hit rates and false alarm rates. The two averages were then combined by an equal weight to obtain the accuracies for each parameter setting. The results are shown on Fig. 3, with the highest accuracy achieved with $C=0.0313$ and $\gamma=0.0313$.

Using the above parameter values, the phonation type classifier was trained with an equal number of regular and irregular tokens. This training set contained all the 1403 irregular vowels and 1403 regular vowels randomly selected from the 8196 available.

5 Evaluation

The subset of the TIMIT test set that was annotated by phonation type was used to evaluate our classifier. It contained 348 irregular and 2680 regular tokens.

The proposed system achieved a 98.85% hit rate (correct recognition of irregular phonation) and a 3.47% false alarm rate (incorrectly classifying regular phonation as irregular). Compared to [1], this is a 7.60% increase in hit rate and a 1.51% decrease in false alarm rate. Note that our results can be directly compared to the results in [1], as the two systems were developed and evaluated using the same training and test sets, and they were designed with the same specifications (the input is a vowel and the output is a binary regular/irregular decision).

Our system produces a higher hit rate and lower false alarm rate than some other, earlier phonation type classifiers [6,7,9]. However, it is not possible to draw conclusions from these comparisons due to differences in the speech corpus used or in the specifications.

6 Conclusions

We proposed a phonation type classifier that aims to distinguish irregularly phonated vowels from regularly phonated ones. Our system achieved a 7.60% higher hit rate and a 1.51% lower false alarm rate than a comparable previously published system. The improvements are due to using a wider range of acoustic cues (integrating cues employed earlier in different systems) and to refining these cues either in terms of the calculation algorithm itself or in terms of the parameters of the algorithm.

The high hit rate (98.85%) and reasonably low false alarm rate (3.47%) are likely to be sufficient for most applications. We have to note however that this classifier has only been tested on vowels. In many cases, this may not limit practical application (e.g. for contributing to the automatic recognition of the emotional state or the identity of the speaker). But in other cases, it would be more useful to have a classifier capable of working on both vowels and consonants, and without phone-level segmentation. Future work should address these issues.

References

1. Surana, S., Slifka, J.: Acoustic cues for the classification of regular and irregular phonation. In: Interspeech 2006, pp. 693–696 (2006)
2. Slifka, J.: Irregular phonation and its preferred role as cue to silence in phonological systems. In: XVIth International Congress of Phonetic Sciences, pp. 229–232 (2007)
3. Henton, C.G., Bladon, A.: Creak as a sociophonetic marker. In: Hyman, L.M., Li, C.N. (eds.) Language, speech and mind: Studies in honour of Victoria A. Fromkin, pp. 3–29. Routledge (1987)

4. Gobl, C., Ní Chasaide, A.: The role of voice quality in communicating emotion, mood and attitude. Speech Communication 40, 189–212 (2003)
5. Surana, K.: Classification of vocal fold vibration as regular or irregular in normal voiced speech. MEng. thesis. MIT (2006)
6. Ishi, C.T., Sakakibara, K.-I., Ishiguro, H., Hagita, N.: A method for automatic detection of vocal fry. IEEE Tr. on Audio, Speech and Language Proc. 16(1), 47–56 (2008)
7. Vishnubhotla, S., Espy-Wilson, C.: Detection of irregular phonation in speech. In: XVIth International Congress of Phonetic Sciences, pp. 2053–2056 (2007)
8. Yoon, T.-J., Zhuang, X., Cole, J., Hasegawa-Johnson, M.: Voice quality dependent speech recognition. In: International Symposium on Linguistic Patterns in Spontaneous Speech (2006)
9. Kiessling, A., Kompe, R., Niemann, H., Nöth, E., Batliner, A.: Voice source state as a source of information in speech recognition: Detection of laryngealizations. In: Rubio-Ayuso, Lopez-Soler (eds.) Speech Recognition and Coding: New advances and trends, pp. 329–332. Springer, Heidelberg (1995)
10. Fawcett, T.: An introduction to ROC analysis. Pattern Recognition Letters 27, 86–874 (2006)
11. Boersma, P.: Accurate short-term analysis of the fundamental frequency and the harmonics-to-noise ratio of a sampled sound. IPS Proceedings 17, 97–100 (1993)
12. Kochanski, G., Grabe, E., Coleman, J., Rosner, B.: Loudness predicts prominence: Fundamental frequency lends little. JASA 118(2), 1038–1054 (2005)
13. Bennett, K.P., Campbell, C.: Support vector machines: Hype or hallelujah? SIGKDD Explorations 2(2), 1–13 (2000)

The Hartley Phase Spectrum as an Assistive Feature for Classification

Ioannis Paraskevas[1] and Maria Rangoussi[2]

[1] Department of Technology of Informatics and Telecommunications,
Technological Educational Institute (T.E.I.) of Kalamata / Branch of Sparta,
7, Kilkis Str., 23100 Sparta, Greece
ioannis.paraskevas21@gmail.com
[2] Department of Electronics,
Technological Educational Institute (T.E.I.) of Piraeus,
250, Petrou Ralli & Thivon Str., 12244 Aigaleo-Athens, Greece
mariar@teipir.gr

Abstract. The phase of a signal conveys critical information for feature extraction. In this work is shown that for certain speech and audio classes where their magnitude content underperforms in terms of recognition rate, the combination of magnitude with phase related features increases the classification rate compared to the case where only the magnitude content of the signal is used. However, signal phase extraction is not a straightforward process, mainly due to the discontinuities appearing in the phase spectrum. Hence, in the proposed method, the phase content of the signal is extracted via the Hartley Phase Spectrum where the sources of phase discontinuities are detected and overcome, resulting in a phase spectrum in which the number of discontinuities is significantly reduced.

Keywords: Fourier Phase Spectrum, Hartley Phase Spectrum, frequency domain feature extraction.

1 Introduction

Accurate phase information extraction is critical for the success of the subsequent speech / audio processing steps; yet difficulties in processing the phase spectrum of speech / audio signals [1] have led researchers to focus their investigation to the magnitude related features. The calculation of the magnitude spectrum preserves the information related to the absolute value of the real and imaginary Fourier components of a signal, but it does not preserve the information related to their signs and changes of signs, [2]. Hence, recently researchers in music processing [3], speech / word recognition [4], [5] and speech processing [6], [7] emphasize the usefulness of phase and propose the use of the phase spectrum.

The majority of the frequency / quefrency features extracted from speech and audio signals are based on the cepstral analysis [8] and they encapsulate the signal's magnitude content (e.g. Mel-Frequency Cepstral Coefficients [9]). Although in most cases the magnitude related features are adequate in terms of their recognition performance,

J. Solé-Casals and V. Zaiats (Eds.): NOLISP 2009, LNAI 5933, pp. 51–59, 2010.
© Springer-Verlag Berlin Heidelberg 2010

there are certain classes of speech and audio signals where the magnitude related features underperform. Hence, this work is focused on the usefulness of the phase spectrum for the pattern recognition of speech and audio signals.

The shape of the Fourier Phase Spectrum for speech and audio signals is characterized by rapid changes appearing as discontinuities; the difficulty in processing the phase spectrum of signals, due to the aforementioned discontinuities, led the researchers to extract spectral features mainly from the magnitude spectrum and thus, ignoring the phase related content of the signals. In order to avoid the aforementioned difficulties in processing the Fourier Phase Spectrum, the Hartley Phase Spectrum is introduced as a useful alternative.

In this work is shown that for certain speech and audio classes, where the magnitude related content of the signals does not form a discriminative feature set for classification, the recognition performance is improved when the magnitude spectral feature set is combined with features extracted from the phase content (specifically the Hartley Phase Spectrum) of the signals. The proposed method is tested on four classes of speech signals [10] and four classes of audio signals [11] - these classes are selected based on the recognition performance of their magnitude content. In the experimental part, the magnitude related and the phase related content of the signals are compared, in terms of their recognition performance, without using a-priori knowledge features e.g. Mel-Frequency Cepstral Coefficients for the speech signals. The recognition performance of the magnitude and the phase related information streams is compared via a Mahalanobis [12] distance-metric classifier. Summarizing, the novelty introduced in this work is the use of the Hartley Phase Spectrum as an assistive feature for the classification of speech and audio signals.

In section 2, the properties of the Hartley Phase Spectrum are established and its applications are presented. In sections 3 and 4, the method for the use of the Hartley Phase Spectrum as an assistive feature for recognition, the combinatory classification scheme and the experimental results are discussed. Summarizing, in section 5 the conclusions of this work are provided.

2   The Hartley Phase Spectrum

In this section, the properties of the Hartley Phase Spectrum, compared to the Fourier Phase Spectrum, are established and the applications of the Hartley Phase Spectrum in speech coding, signal localization and signal analysis, are presented.

The Fourier Phase Spectrum (FPS) is defined as:

$$\varphi(\omega) = \arctan\left(\frac{S_I(\omega)}{S_R(\omega)}\right) \tag{1}$$

where $-\pi \leq \varphi(\omega) < \pi$, $S_R(\omega)$ and $S_I(\omega)$ are the real and the imaginary components of $S(\omega)$, respectively and arctan denotes the inverse tangent function.

The FPS is characterized by rapid changes, appearing as discontinuities, which are categorized as: i) 'extrinsic' discontinuities: caused due to the use of the inverse tangent function which results in 'wrapping' ambiguities [13] and ii) 'intrinsic' discontinuities: caused due to properties of the signal itself [14]. However, the heuristic approach of the 'unwrapping' algorithm, which is used in order to compensate the extrinsic discontinuities appearing across the FPS, introduces ambiguities that are more severe in the presence of even the lower noise level [15].

It is important to mention that an alternative approach for the evaluation of the FPS, without suffering from the extrinsic discontinuities, is via the z-transform [16]. Indeed, although computation of the phase component of each individual 'zero' (root of the polynomial formed by the signal) still employs the inverse tangent function, phase values of individual 'zeros' are always constrained within $\pm\pi$ radians. Hence, phase is not wrapped around zero and consequently wrapping ambiguities do not arise (see [17] for a similar approach applied to speech utterances). However, calculation of the roots ('zeros') of a high degree polynomial (speech / audio signal) is sought by numerical methods that often produce inaccuracies in the values of the roots which increase with polynomial order [18] and therefore, this method for the evaluation of the FPS is not used in this work.

To overcome the aforementioned difficulties in evaluating the phase spectrum, the Hartley Phase Spectrum (HPS) is introduced. The HPS is defined as the division of the Hartley Transform [19] over the Fourier Magnitude Spectum,

$$Y(\omega) = \frac{H(\omega)}{M(\omega)} = \frac{M(\omega)(\cos(\varphi(\omega)) + \sin(\varphi(\omega)))}{M(\omega)} = \cos(\varphi(\omega)) + \sin(\varphi(\omega)) \qquad (2)$$

where $H(\omega)$ denotes the Hartley Transform and $M(\omega)$, $\varphi(\omega)$ the Fourier Magnitude Spectrum and the FPS, respectively. The HPS, also known as the 'scaled' or Whitened Hartley Spectrum, is a function of the Fourier Phase $\varphi(\omega)$, [20].

The HPS, unlike the FPS, does not suffer from the extrinsic discontinuities. Specifically, for the Hartley Phase case, the phase spectral function is not wrapped around zero and consequently, the unwrapping algorithm is not required. Similarly though to the FPS, the HPS conveys the intrinsic discontinuities, as they are caused due to certain properties of the signal. However, the intrinsic discontinuities can be easily compensated from the HPS unlike the case of its close relative the Whitened Fourier Spectrum (WFS), where there is no known method to overcome them [15].

Moreover, the HPS, due to its structure, is less affected by the presence of noise; the noise immunity of the HPS, compared to the FPS, is justified via the shapes of the respective Probability Density Functions [21]. The advantages of the HPS carry over to the Hartley Phase Cepstrum, thanks to the analytic relations holding between the spectral and the cepstral domains, [15], [21], [22].

2.1 Applications of the Hartley Phase Spectrum

The useful properties of the HPS have already found application in:

i)   Speech coding [20]. The HPS is a real function which is bounded between $\pm\sqrt{2}$ – a property which is particularly useful for coding applications.

ii) Signal localization [22]. The Fourier Phase Cepstrum (FPC), unlike the Hartley Phase Cepstrum (HPC), cannot yield the location of more than a single pulse due to the heuristic and non-invertible nature of the unwrapping algorithm which is applied in the FPS.

iii)   Model-based signal analysis [21]. The noise robustness of the HPS, compared
to the FPS, is shown via the shape of the HPS's Probability Density Function
(PDF) for the case of noisy signals. The noise robustness of the HPS is due to
the lack of wrapping ambiguities.

3   The Hartley Phase Spectrum as an Assistive Tool for Feature Extraction and Classification

In this section, the method for the use of the Hartley Phase Spectrum as an assistive
feature for classification, is described. Specifically, in the following subsections, the
implementation of the spectrograms, the features extracted from each spectrogram,
the classification scheme employed and the databases where the proposed method is
tested, are described.

3.1   Spectrograms' Implementation and Feature Selection

Each speech and audio signal is divided into equal-length frames (256 samples) with
zero-padding of the last frame if necessary, windowed with a Hanning window and
transformed to the frequency domain. The transformed frames are placed row-wise in
a matrix and hence the spectrograms are implemented.

The five spectrograms that are implemented in this subsection are compared in
terms of their recognition performance. The experimental results show that for certain
speech and audio classes where their magnitude related content underperforms in
terms of recognition, the use of their phase related content, as an assistive feature,
increases the classification rate (section 4).

The implementation of the spectrograms is based on the following formulas of
their corresponding spectra:

i)   Fourier Magnitude Spectrum (FMS),

$$M(\omega) = \sqrt{S_R^{\,2}(\omega) + S_I^{\,2}(\omega)} \qquad (3)$$

where $S_I(\omega)$ and $S_R(\omega)$ denote the imaginary and the real part, respectively, of the
complex Fourier spectrum of the signal, $S(\omega)$.

ii)   Hartley Magnitude Spectrum (HMS),

$$N(\omega) = \sqrt{\left|H(\omega)H^*(\omega)\right|} = M(\omega)\sqrt{\left|\cos(2\varphi(\omega))\right|} \qquad (4)$$

where $H(\omega) = M(\omega)[\cos(\varphi(\omega)) + \sin(\varphi(\omega))]$ denotes the Hartley Transform and
$H^*(\omega) = M(\omega)[\cos(\varphi(\omega)) - \sin(\varphi(\omega))]$ denotes the 'quasi-conjugate' Hartley Transform. Hence, the HMS is a combination of the FMS, $M(\omega)$ with the FPS, $\varphi(\omega)$, of
the signal.

iii)   Fourier Phase Spectrum (FPS), see equation (1).
iv)   Hartley Phase Spectrum (HPS), see equation (2).
v)   Hartley Transform (HT),

$$H(\omega) = M(\omega)[\cos(\varphi(\omega)) + \sin(\varphi(\omega))] \qquad (5)$$

Each of the five aforementioned spectrograms has to be presented to the classifier with its dimensionality reduced. Hence, statistical features are calculated from each spectrogram for each signal, in order to compress the information into a compact feature vector. The eight statistical features, which form the feature vector, are chosen empirically as the most representative descriptive statistics, namely: the variance, the skewness, the kurtosis, the entropy, the inter-quartile range, the range (note that the range is not used for the HPS case, because it is a bounded function), the median and the mean absolute deviation. These statistical features form an [8 x 1] - sized feature vector.

Each signal is classified using a Mahalanobis distance [12] metric classifier. The Mahalanobis distance is calculated between each test vector and each reference vector. The test vector is matched to the class (reference vector) closest to it [23].

3.2 Databases

The two databases used for the experiments consist of certain (selected) classes of speech and audio signals that their recognition performance obtained based on their magnitude related features is low. Hence,

- the four selected classes of speech signals are: /h/ e.g. in 'hill', /s/ e.g. in 'sister', /f/ e.g. in 'fat' and /th/ e.g. in 'thin' (speech database: TIMIT [10]) and
- the four selected classes of audio signals are: firing a .22 caliber handgun, firing a World War II German rifle, firing a cannon and firing a pistol (audio database: 505 Digital Sound Effects [11]).

For both databases, each class consists on average of ten signal recordings where seven of them are used as training data and the rest as test data.

3.3 Spectrogram Selection and Classification Scheme

The five spectrograms described in subsection 3.1 can be categorized in three groups, based on the information they convey.

- Category 1 - spectrograms that convey purely the signal's magnitude content: In this category belongs the Fourier Magnitude (equation (3)) spectrogram which preserves only the magnitude content of the signal.
- Category 2 - spectrograms that convey purely the signal's phase content: In this category belong the Hartley Phase (equation (2)) spectrogram and the Fourier Phase (equation (1)) spectrogram that preserve only the phase content of the signal.
- Category 3 - spectrograms that convey a combination of the signal's magnitude and phase content: In this category belong the Hartley Magnitude (equation (4)) spectrogram and the Hartley Transform (equation (5)) spectrogram.

For the experimental part (section 4), one spectrogram is selected from each of the three aforementioned categories in order to avoid information overlapping.

- From Category 1, the Fourier Magnitude spectrogram is selected.
- In Category 2, the Hartley Phase spectrogram performs better in terms of recognition compared to the Fourier Phase spectrogram (6.5% higher classification rate

on average for both databases). The lower classification rate of the Fourier Phase spectrogram is due to the ambiguities introduced by the phase unwrapping. Hence, from Category 2, the Hartley Phase spectrogram is selected due to its recognition performance.

- In Category 3, the Hartley Magnitude spectrogram performs better, in terms of its recognition rate, compared to the spectrogram implemented based on the Hartley Transform (10.0% higher classification rate on average for both databases). Therefore, the Hartley Magnitude spectrogram is preferred.

Hence, the three information streams (spectrograms) selected are: the Fourier Magnitude spectrogram, the Hartley Phase spectrogram and the Hartley Magnitude spectrogram. As will be presented in section 4, the feature vectors extracted from the selected spectrograms are combined (Figure 1) and the classification rate obtained is compared with the classification rate of the magnitude related information stream i.e. the Fourier Magnitude spectrogram.

The classification scheme which combines the three aforementioned spectrograms is based on the 'majority vote' decision rule, as presented in Figure 1.

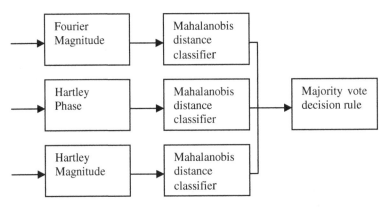

Fig. 1. Majority vote decision rule. A signal is classified to a certain class if two or more of the three feature vectors vote for this class.

Summarizing, for each signal three feature vectors are formed which are extracted from the three aforementioned spectrograms. A speech / audio signal is classified based on the majority vote decision rule (a Mahalanobis distance metric classifier is used for each information stream). Thus, in case two or three out of the three independent 'experts' (feature vectors) agree, then a signal is classified to this class. In case of tie, the decision is taken based on the Fourier Magnitude independent expert.

4   Experimental Results and Discussion

In this section, the aforementioned classification scheme is tested on the speech database and on the audio database and the experimental results are presented. Moreover, additional observations related to the efficient feature extraction from the phase spectrum, are discussed.

4.1 Classification Results for Speech and Audio Signals

The classification rate of the Fourier Magnitude feature vectors is compared with the classification rate of the proposed combinatory scheme. The speech database and the audio database used for the aforementioned comparison, consist - each of them - of four pre-selected classes where the Fourier Magnitude feature vectors underperform in terms of their recognition performance (subsection 3.2).

Hence, for the speech signals (second row of Table 1), the overall classification rate reaches 75.0% in case all three information streams are combined together (combinatory scheme, subsection 3.3) whereas, the classification rate is 52.8% in case only the Fourier Magnitude information stream is used. Similarly, for the audio signals (third row of Table 1), the overall classification rate reaches 93.0% in case all three information streams are combined together (combinatory scheme, subsection 3.3) whereas, the classification rate is 51.7% in case only the Fourier Magnitude information stream is used. Thus, for both databases, the classification rate becomes higher when the magnitude-related feature vectors are used together with the phase-related feature vectors (combinatory scheme), compared to the case where only the magnitude-related (Fourier Magnitude spectrogram) features vectors are used.

Table 1. Correct classification scores (%) for speech signals (2nd row) and for audio signals (3rd row)

Feature vectors extracted from:	Fourier Magnitude Spectrogram	Fourier Magnitude, Hartley Phase & Hartley Magnitude Spectrograms (combinatory scheme – Figure 1)
Average scores for speech signals (4 classes)	52.8	75.0
Average scores for audio signals (4 classes)	51.7	93.0

The experimental results show that the phase spectrum of the speech and audio signals conveys useful information. Therefore, for certain speech and audio classes where the magnitude content of the signals cannot form an efficient discriminative feature set, their phase related content forms an assistive feature set for classification.

4.2 Additional Classification Results and Discussion

In this subsection, useful observations related to the efficient feature extraction from the phase spectrum of speech and audio signals, are discussed. Specifically, as indicated by the experimental results, the information content of the phase spectrum is more efficiently encapsulated in the feature vectors, when:

i)   the discontinuities of the phase spectrum are compensated [24]. Specifically, there is a 24.9% increase in the classification rate (average for both databases) when the intrinsic discontinuities are compensated from the Hartley Phase spectrogram (no extrinsic discontinuities exist in the HPS).

The same observation holds for the Fourier Phase case where the highest classification rate is reached when both the extrinsic and the intrinsic discontinuities are compensated. Thus, the experimental results indicate that the discontinuities appearing in the phase spectrum affect the recognition performance and therefore, the feature vectors encapsulate the phase content more efficiently when the discontinuities are compensated.

ii) the feature vectors are extracted from the 'difference' of the phase spectrogram (i.e., the phase spectrogram matrix is replaced by a matrix formed by taking the first-order discrete difference along each row of the original matrix) [24]. Specifically, for the Hartley Phase spectrogram, the recognition rate is improved by 14.0% (average for both databases) when the phase difference is calculated.

The aforementioned observation also holds for the Fourier Phase case and indicates that the phase difference rather than the phase itself is a more informative feature for classification. This performance improvement is in agreement with results reported in [25], claiming that phase difference is less affected by noise than phase itself.

Therefore, based on both the aforementioned observations, the feature vectors of the Hartley Phase spectrogram used in the combinatory scheme (Figure 1, Table 1) are extracted after the discontinuities are compensated and the first-order discrete difference along each row of the Hartley Phase spectrogram is evaluated.

5   Conclusions

This work, focused on the feature extraction stage of the pattern recognition process, shows the importance of the phase spectrum as a feature extraction tool for classification. The experimental results show that for certain speech and audio classes where the magnitude features underperform in terms of their recognition rate, the combination of magnitude with phase related features increases the classification score. Thus, the experimental results indicate that the phase spectrum and specifically the Hartley Phase Spectrum conveys useful information, which is used as an additional / assistive feature for classification.

References

1. Alsteris, L.D., Paliwal, K.K.: Further intelligibility results from human listening tests using the short-time phase spectrum. Speech Communication 48, 727–736 (2006)
2. McGowan, R., Kuc, R.: A direct relation between a signal time series and its unwrapped phase. IEEE Transactions on Acoustics, Speech, and Signal Processing 30(5), 719–726 (1982)
3. Eck, D., Casagrande, N.: Finding meter in music using an autocorrelation phase matrix and Shannon entropy. In: Proc. of the 6th Int. Conference on Music Information Retrieval (ISMIR), UK, pp. 312–319 (2005)

4. Schlüter, R., Ney, H.: Using phase spectrum information for improved speech recognition performance. In: Proc. of the Int. Conference on Acoustics, Speech, and Signal Processing (ICASSP), USA, vol. 1, pp. 133–136 (2001)
5. Alsteris, L.D., Paliwal, K.K.: Evaluation of the modified group delay feature for isolated word recognition. In: Proc. of the 8th International Symposium on Signal Processing and its Applications (ISSPA), Australia, pp. 715–718 (2005)
6. Paliwal, K.K., Alsteris, L.D.: On the usefulness of STFT phase spectrum in human listening tests. Speech Communication 45, 153–170 (2005)
7. Bozkurt, B., Couvreur, L., Dutoit, T.: Chirp group delay analysis of speech signals. Speech Communication 49, 159–176 (2007)
8. Furui, S.: Cepstral analysis technique for automatic speaker verification. IEEE Transactions on Acoustics, Speech, and Signal Processing 29(2), 254–272 (1981)
9. Davis, S., Mermelstein, P.: Comparison of parametric representations for monosyllabic word recognition in continuously spoken sentences. IEEE Transactions on Acoustics, Speech, and Signal Processing 28(4), 357–366 (1980)
10. TIMIT® Acoustic-Phonetic Continuous Speech Corpus (1993)
11. Audio Database: 505 Digital Sound Effects. (Disk 3/5: 101 Sounds of the Machines of War), Delta (1993)
12. Mahalanobis, P.C.: On the generalized distance in statistics. Proceedings of the National Institute of Science of India 12, 49–55 (1936)
13. Tribolet, J.: A new phase unwrapping algorithm. IEEE Transactions on Acoustics, Speech and Signal Processing 25(2), 170–177 (1977)
14. Al-Nashi, H.: Phase Unwrapping of Digital Signals. IEEE Transactions on Acoustics, Speech and Signal Processing 37(11), 1693–1702 (1989)
15. Paraskevas, I., Rangoussi, M.: The Hartley Phase Cepstrum as a Tool for Signal Analysis. In: Chetouani, M., Hussain, A., Gas, B., Milgram, M., Zarader, J.-L. (eds.) NOLISP 2007. LNCS (LNAI), vol. 4885, pp. 204–212. Springer, Heidelberg (2007)
16. Proakis, J.G., Manolakis, D.G.: Digital Signal Processing Principles, Algorithms, and Applications, ch. 4, 5. Macmillan Publishing Company, Basingstoke (1992)
17. Bozkurt, B., Dutoit, T.: Mixed-phase speech modeling and formant estimation, using differential phase spectrums. In: Proc. Voice Quality: Functions, Analysis and Synthesis (VOQUAL), Switzerland, pp. 21–24 (2003)
18. Sitton, G.A., Burrus, C.S., Fox, J.W., Treitel, S.: Factoring very-high-degree polynomials. IEEE Signal Processing Magazine 6, 27–42 (2003)
19. Bracewell, R.N.: The Fourier Transform and Its Applications, ch. 19. McGraw-Hill Book Company, New York (1986)
20. Chilton, E.: An 8kb/s speech coder based on the Hartley transform. In: Proc. Communication Systems: Towards Global Integration (ICCS), Singapore, vol. 1, pp. 13.5.1–13.5.5 (1990)
21. Paraskevas, I., Rangoussi, M.: The Hartley Phase Spectrum as a noise-robust feature in speech analysis. In: Proc. of the ISCA Tutorial and Research Workshop (ITRW) on Speech Analysis and Processing for Knowledge Discovery, Denmark (2008)
22. Paraskevas, I., Rangoussi, M.: The Hartley phase cepstrum as a tool for improved phase estimation. In: Proc. of the 16th Int. Conference on Systems, Signals and Image Processing (IWSSIP), Greece (2009)
23. Webb, A.R.: Statistical Pattern Recognition, 2nd edn., ch. 9. John Wiley & Sons, Ltd., Chichester (2002)
24. Paraskevas, I., Chilton, E., Rangoussi, M.: Audio Classification Using Features Derived from The Hartley Transform. In: Proc. of the 13th Int. Conference on Systems, Signals and Image Processing (IWSSIP), Hungary, pp. 309–312 (2006)
25. Gough, P.: A particular example of phase unwrapping using noisy experimental data. IEEE Transactions on Acoustics, Speech, and Signal Processing 31(3), 742–744 (1983)

Speech Enhancement for Automatic Speech Recognition Using Complex Gaussian Mixture Priors for Noise and Speech

Ramón F. Astudillo, Eugen Hoffmann, Philipp Mandelartz,
and Reinhold Orglmeister

Department of Energy and Automation Technology, TU-Berlin,
Einsteinufer 17, 10587 Berlin, Germany
ramon@astudillo.com, Eugen.Hoffmann.1@tu-berlin.de,
reinhold.orglmeister@tu-berlin.de
http://www.emsp.tu-berlin.de

Abstract. Statistical speech enhancement methods often rely on a set of assumptions, like gaussianity of speech and noise processes or perfect knowledge of their parameters, that are not fully met in reality. Recent advancements have shown the potential improvement in speech enhancement obtained by employing supergaussian speech models conditioned on the estimated signal to noise ratio. In this paper we derive a supergaussian model for speech enhancement in which both speech and noise priors are assumed to be complex Gaussian mixture models. We introduce as well a method for the computation of the noise prior based on the noise variance estimator used. Finally, we compare the developed estimators with the conventional Ephraim-Malah filters in the context of robust automatic speech recognition.

Keywords: Supergaussian Priors, GMM, IMCRA, MMSE-STSA, MMSE-LSA.

1 Introduction

In the standard framework for enhancement of speech corrupted by additive noise, a digitalized noisy signal $y(t) = x(t) + d(t)$ is considered as containing clean speech $x(t)$ and noise $d(t)$. This signal is usually transformed into the time-frequency domain, where modeling of speech and noise is easier, by employing the short-time Fourier transform (STFT). Let $Y = X + D$ be a given Fourier coefficient of the STFT of $y(t)$ and let X and D be the corresponding Fourier coefficients of the clean speech and noise signals. Statistical speech enhancement methods obtain an estimation $\hat{X}$ of each clean Fourier coefficient X, from each noisy coefficient Y by employing conventional estimation methods like maximum likelihood (ML), maximum a posteriori (MAP) or minimum mean square error (MMSE). Such methods require establishing a priori probability distributions for the speech and noise Fourier coefficients $p(X), p(D)$. Each estimation $\hat{X}$ is normally expressed in form of a gain function or "filter" applied

J. Solé-Casals and V. Zaiats (Eds.): NOLISP 2009, LNAI 5933, pp. 60–67, 2010.

to each time-frequency element as $\hat{X} = G \cdot Y$, where G is determined by the a priori distributions chosen, their estimated statistical parameters and the value of Y. Once an estimation of each clean Fourier coefficient has been obtained, the time-domain clean signal $\hat{x}(t)$ can be recovered by applying the inverse Fourier transform (ISTFT).

A majority of the speech enhancement methods employed in STFT domain, like power subtraction, Wiener [1] and Ephraim-Malah filters [2,3], use the so called Gaussian model. Under this model the distributions of X and D are assumed to be zero mean, circular, and complex Gaussian. The assumption of Gaussianity has been however challenged by various authors. Supergaussian a priori distributions for speech, like those employing Laplace or Gamma [4,6], generalized Gamma [8] or Rayleigh mixture model [9] distributions, have been demonstrated to offer a better fit to the empirically obtained distribution of the speech Fourier coefficients. MAP [6] and MMSE [4,8] estimators derived under this models have also shown to yield better speech enhancement properties in terms of SNR.

This paper studies the use of Gaussian mixture models for speech enhancement in automatic speech recognition (ASR) applications. The case considered here is the one in which both speech and noise a priori distributions are assumed to be complex Gaussian mixture models and it can be considered a generalization of the Rayleigh mixture speech prior model in [9]. The paper is divided as follows. Section two develops a method for the computation of the noise prior based on the noise variance estimation method used. In section three, the posterior distribution for the complex Gaussian mixture model, which corresponds to a Rice mixture model, is derived. From this posterior, MMSE estimators of the Short-Time Spectral Amplitude (MMSE-STSA) and Log-Spectral-Amplitude (MMSE-LSA), equivalent to the Ephraim-Malah filters under the Gaussian model, are also derived. In section four the efficiency of the proposed supergaussian noise estimator and its combination with the Rayleigh mixture model estimator derived in [9] is examined in the context of robust ASR. Finally, section five outlines the conclusions.

2   A Gaussian Mixture Model for Noise

Under the conventional Gaussian model, the probability density function (PDF) of a complex valued noise Fourier coefficient $D = D_R + jD_I$ can be obtained as the product of the PDFs of its real an imaginary components. These PDFs are zero mean and Gaussian distributed with variance σ_D^2 leading to the following expression for the PDF

$$p(D) = p(D_R)p(D_I) = \frac{1}{\pi\lambda_D} \exp\left(-\frac{|D|^2}{\lambda_D}\right), \tag{1}$$

where $\lambda_D = 2\sigma_D^2$ corresponds to the variance of D. As stated by Ephraim and Cohen [7] the Gaussian assumption is a convenient simplification since the a priori distribution is, in fact, conditioned on the knowledge of its variance. In the case of the noise prior distribution considered here, the PDF of the real component D_R of a noise Fourier coefficient, can be expressed as as the following continuous Gaussian mixture model [7]

$$p(D_R) = \int p(D_R|\sigma_D^2)p(\sigma_D^2)d\sigma_D^2, \tag{2}$$

where $p(D_R|\sigma_D^2)$ is Gaussian and consequently the selection of $p(\sigma_D^2)$ determines the distribution of the prior $p(D_R)$.

Based on this general model we propose the following complex Gaussian mixture model as prior distribution for noise,

$$p(D) = \sum_{j=1}^{J} w_{D_j} \frac{1}{\pi \hat{\lambda}_{D_j}} \exp\left(-\frac{|D|^2}{\hat{\lambda}_{D_j}}\right), \tag{3}$$

where the mixture weights w_{D_j} are normalized to sum up to one and the noise variance of each mixture component λ_{D_j} is normally not available and has been replaced by its estimation $\hat{\lambda}_{D_j}$. If we now integrate Eq. 3 over the imaginary part D_I of D, the resulting distribution of D_R corresponds to the following discrete Gaussian mixture model

$$p(D_R) = \int p(D)dD_I = \sum_{j=1}^{J} w_{D_j} \frac{1}{\sqrt{2\pi\hat{\sigma}_{D_j}^2}} \exp\left(-\frac{D_R^2}{2\hat{\sigma}_{D_j}^2}\right), \tag{4}$$

If we interpret each Gaussian mixture as the following conditional distribution

$$p(D_R|\hat{\sigma}_D^2 = \hat{\sigma}_{D_j}^2) = \frac{1}{\sqrt{2\pi\hat{\sigma}_{D_j}^2}} \exp\left(-\frac{D_R^2}{2\hat{\sigma}_{D_j}^2}\right), \tag{5}$$

we can consider the discrete mixture model in Eq. 4 as an approximation of the continuous mixture model proposed by Ephraim and Cohen and given in Eq. 2

$$p(D_R) = \int p(D_R|\hat{\sigma}_D^2)p(\hat{\sigma}_D^2)d\hat{\sigma}_D^2 \approx \sum_{j=1}^{J} w_{D_j} p(D_R|\hat{\sigma}_D^2 = \hat{\sigma}_{D_j}^2), \tag{6}$$

where in our case

$$w_{D_j} = p(\hat{\sigma}_D^2 = \hat{\sigma}_{D_j}^2), \tag{7}$$

with $j = 1 \cdots J$, is a discretization of the distribution of the noise variance estimate $p(\hat{\sigma}_D^2)$. Any speech enhancement system must include a method for the estimation of the variance of noise $\lambda_D = 2\sigma_D^2$. For a given method, it is possible to obtain samples of the noise estimate $\hat{\sigma}_D^2$ from which $p(\hat{\sigma}_D^2 = \hat{\sigma}_{D_j}^2)$ can be approximated[1]. For a given discrete approximation of $p(\hat{\sigma}_D^2)$ the weights w_{D_j} and the normalized mixture variances

---

[1] Note that unlike D, $\hat{\sigma}_D^2$ can be approximated directly through histograms since it can be considered an ergodic process.

$$r_{D_j} = \frac{\hat{\sigma}^2_{D_j}}{\sum_{j=1}^{J} w_{D_j} \hat{\sigma}^2_{D_j}} = \frac{\hat{\lambda}_{D_j}}{\sum_{j=1}^{J} w_{D_j} \hat{\lambda}_{D_j}}, \tag{8}$$

determine the form of the complex Gaussian mixture model noise prior $p(D)$ given in Eq. 3. This allows us to determine a prior distribution for noise for a given noise variance estimation method.

3   Derivation of MMSE Clean Speech Estimators Using Gaussian Mixture Priors

Let us consider a general Gaussian mixture model for speech enhancement in which both noise and speech a priori distributions are modeled as complex Gaussian mixture models. The PDF of the clean speech Fourier coefficient corresponds then to

$$p(X) = \sum_{i=1}^{I} w_{X_i} \frac{1}{\pi \lambda_{X_i}} \exp\left(-\frac{|X|^2}{\lambda_{X_i}}\right), \tag{9}$$

where equivalently to Eq. 8 we define the ratios

$$r_{X_i} = \frac{\lambda_{X_i}}{\sum_{i=1}^{I} w_{X_i} \lambda_{X_i}} \tag{10}$$

and the PDF for the noise Fourier coefficient $p(D)$ was defined in Eq. 3.

In order to derive a minimum mean square error (MMSE) estimator, it is necessary to determine first the form of the likelihood distribution $p(Y|X)$. Under the assumption of additive noise, the likelihood distribution corresponds to the noise prior proposed in Eq. 3 but centered around X

$$p(Y|X) = \sum_{j=1}^{J} w_{D_j} \frac{1}{\pi \hat{\lambda}_{D_j}} \exp\left(-\frac{|Y - X|^2}{\hat{\lambda}_{D_j}}\right). \tag{11}$$

MMSE-STSA and MMSE-LSA estimators use the fact that the phase α of the clean Fourier coefficient $X = Ae^{j\alpha}$ is an uniform distributed nuisance parameter that can be integrated out from the model. Since this also holds for the Gaussian mixture model, it is possible to marginalize the phase from Eq. 9 thus obtaining a Rayleigh mixture model

$$p(A) = \int_0^{2\pi} p(Ae^{j\alpha}) A d\alpha = \sum_{i=1}^{I} w_{X_i} \frac{2A}{\lambda_{X_i}} \exp\left(-\frac{A^2}{\lambda_{X_i}}\right). \tag{12}$$

The phase α can also be integrated out from the likelihood function in Eq. 11, in a similar way as in [1], leading to

$$p(Y|A) = \sum_{j=1}^{J} w_{D_j} \frac{1}{\pi \hat{\lambda}_{D_j}} \exp\left(-\frac{|Y|^2 + A^2}{\hat{\lambda}_{D_j}}\right) I_0\left(\frac{2|Y|A}{\hat{\lambda}_{D_j}}\right), \tag{13}$$

where I_0 corresponds to the zeroth order modified Bessel function. We now apply the Bayes theorem to obtain the posterior distribution from Eqs. 12, 13 as

$$p(A|Y) = \frac{p(Y|A)p(A)}{\int p(Y|A)p(A)dA} =$$

$$\sum_{i=1}^{I}\sum_{j=1}^{J} \Omega_{ij} \frac{2A}{\frac{\xi_{ij}}{1+\xi_{ij}} \frac{|Y|^2}{\gamma_j}} \exp\left(-\frac{A^2 + \left(\frac{\xi_{ij}}{1+\xi_{ij}}|Y|\right)^2}{\frac{\xi_{ij}}{1+\xi_{ij}} \frac{|Y|^2}{\gamma_j}}\right) I_0\left(\frac{2A\gamma_j}{|Y|}\right), \tag{14}$$

which corresponds to a Rice mixture model (RiMM) with weights

$$\Omega_{ij} = \frac{\tilde{\Omega}_{ij}}{\sum_{i=1}^{I}\sum_{j=1}^{J}\tilde{\Omega}_{ij}}, \tag{15}$$

where

$$\tilde{\Omega}_{ij} = w_{X_i} w_{D_j} \frac{\gamma_j}{1+\xi_{ij}} \exp\left(-\frac{\gamma_j}{1+\xi_{ij}}\right) \tag{16}$$

and where

$$\xi_{ij} = \frac{\lambda_{X_i}}{\lambda_{D_j}} = \frac{r_{X_i}}{r_{D_j}}\xi, \tag{17}$$

and

$$\gamma_j = \frac{|Y|^2}{\lambda_{D_j}} = \frac{\gamma}{r_{D_j}}, \tag{18}$$

are expressed as a function of the a posteriori and a priori signal to noise ratios, γ and ξ, defined in [2, Eqs. 9, 10].

Since the posterior distribution of the clean speech amplitude given in Eq. 14 is a weighted sum of the posterior of the Ephraim-Malah filters, computing the new estimator gains is trivial, yielding

$$G(\xi_{ij}, \gamma_j)^{\text{RiMM-STSA}} = \frac{E\{A|Y\}}{|Y|} = \sum_{i=1}^{I}\sum_{j=1}^{J} \Omega_{ij} G(\xi_{ij}, \gamma_j)^{\text{STSA}} \tag{19}$$

and

$$G(\xi_{ij}, \gamma_j)^{\text{RiMM-LSA}} =$$

$$\frac{\exp\left(\frac{d}{d\mu}E\{A^\mu|Y\}|_{\mu=0}\right)}{|Y|} = \prod_{i=1}^{I}\prod_{j=1}^{J} \left(G(\xi_{ij}, \gamma_j)^{\text{LSA}}\right)^{\Omega_{ij}}, \tag{20}$$

where $G(\xi_{ij}, \gamma_j)^{\text{STSA}}$ and $G(\xi_{ij}, \gamma_j)^{\text{LSA}}$ correspond to the gains for the MMSE-STSA and MMSE-LSA estimators under the Gaussian model given by [2, Eq. 7] and [3, Eq. 20] respectively, evaluated for each mixture component of the noise and speech priors.

4   Experimental Results

The efficiency of the derived estimators was tested in a robust ASR test. For this purpose, an ASR system was trained using the AURORA5 database resources [10] with the whole clean speech train-set. Tests were carried out using the whole clean speech test-set of the AURORA5 to which white noise and highly instationary wind noise was added at different SNR levels. For the two types of speech enhancement methods derived, MMSE-STSA and MMSE-LSA, three alternatives were tested. These were: Conventional Gaussian prior for speech and noise (GS+GN), Gaussian prior for speech combined with the supergaussian noise prior introduced in this paper (GS+SN) and finally the Rayleigh mixture model speech prior introduced in [9] combined with the supergaussian noise prior introduced in this paper (SS+SN).

The supergaussian noise prior was determined as explained in section two from speech enhancement experiments in which white noise and instationary-noise samples, not present in the test-set, were added to clean speech. The method used for the estimation of $\hat{\sigma}_D^2$ was the improved minima controlled recursive averaging (IMCRA) [5]. From the noise variance samples obtained in the experiments a 100 bin histogram was computed as shown in Fig. 1 (left). Accordingly to the model in Eq. 6, this corresponds to a $J = 100$ mixture prior distribution as in Fig. 1 (right). To reduce the number of mixtures needed for the modeling of the prior, samples from this distribution were drawn using Montecarlo simulation and a $J = 5$ mixture prior distribution was trained with these samples using the conventional expectation-maximization technique. The resulting prior is almost identical to the original 100 mixture prior and overlaps completely with it in Fig. 1 (right). The Rayleigh mixture model parameters were trained using the expectation-maximization method and clean speech samples. These samples were

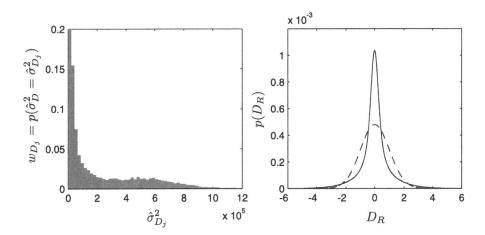

Fig. 1. Left: Approximation of the PDF of $\hat{\sigma}_D^2$ by using a 100 bin histogram. Right, solid: Gaussian mixture model prior for D_R with $J = 100$ mixtures obtained from the histogram on the left through Eqs. 6, 7, 8. Right, dashed: Gaussian prior for D_R.

Table 1. Average word error rates (WER) [%] for the MMSE-STSA and MMSE-LSA estimators using Gaussian (G*) and supergaussian priors (S*) for speech (*S) and noise (*N). Results that improve conventional Gaussian estimators displayed in bold.

Noise Type	SNR	Noisy	MMSE-STSA			MMSE-LSA		
			GS+GN	GS+SN	SS+SN	GS+GN	GS+SN	SS+SN
	∞	1.4	1.4	1.4	1.5	1.5	1.5	1.9
White	10	7.7	3.1	3.3	4.0	3.3	**2.9**	4.2
	05	16.6	5.7	6.4	8.2	5.8	**4.9**	7.6
	00	29.3	12.5	14.3	17.8	13.0	**10.6**	15.6
Wind	05	2.0	2.0	**1.8**	2.2	2.3	**1.9**	2.7
	00	3.9	3.6	**2.7**	3.6	3.8	**2.9**	3.8
	-05	9.9	9.0	**6.8**	**8.4**	8.9	**6.8**	**7.3**

obtained, following the criteria used in [6,9], for a range of ξ of 19dB to 21dB. A value of $I = 5$ mixtures was used.

Results displayed in Table 1 show that MMSE estimators employing only the supergaussian noise prior (GS+SN) reduce the word error rates (WER) under the presence of noise a 10%-25%. However, only the MMSE-LSA estimator is succesful both on white and instationary noises. The use of the Rayleigh mixture prior for speech, in addition to the supergaussian noise models, (SS+SN) had a negative effect in all scenarios and the resulting estimator improves its Gaussian counterpart only for instationary noises at low SNRs.

5   Conclusions

A complex Gaussian mixture model for noise has been introduced along with a method for the estimation of its parameters for an specific noise variance estimator. The posterior distribution along with MMSE-STSA and MMSE-LSA estimators for the general case in which both speech and noise priors are complex Gaussian distributed have been derived. This estimators outperform the conventional estimators in most robust ASR tasks tested when using supergaussian priors for noise. The additional use of supergaussian priors for speech provided no improvement. This does however not contradict the results in [9] since an optimal estimator in terms of SNR must not always be optimal for robust ASR and vice versa. With respect to the presented approach, a limiting factor for efficiency is the fact that Ephraim and Cohen's model given in eq. 2 assumes the conditional distribution $p(D_R|\sigma_D^2)$ to be Gaussian, which might not always hold. This limitation adds to the fact that the prior parameters are obtained in a data driven way, which is a source of potential overfitting problems. Future research should be centered in overcoming those limiting factors. Since, unlike for other supergaussian estimators, a closed-form solution for the posterior distribution is available, another interesting research direction is the combination of the Rice mixture posterior with uncertainty propagation techniques as in [11].

References

1. MCAulay, R.J., Malpass, L.M.: Speech enhancement Using a Soft-Decision Noise Suppression Filter. IEEE Trans. ASSP 28(2), 137–145 (1980)
2. Ephraim, Y., Malah, D.: Speech enhancement using a minimum mean-square error short-time amplitude estimator. IEEE Trans. ASSP 32(6), 1109–1121 (1984)
3. Ephraim, Y., Malah, D.: Speech enhancement using a minimum mean-square error log-spectral amplitude estimator. IEEE Trans. ASSP 33(2), 443–445 (1985)
4. Martin, R.: Speech enhancement using MMSE short time spectral estimation with gamma distributed speech priors. IEEE Int. Conf. Acoustics, Speech, Signal Processing 1, 253–256 (2002)
5. Cohen, I.: Noise Spectrum Estimation in Adverse Environments: Improved Minima controlled Recursive Averaging. IEEE Trans. ASSP 11(5), 466–475 (2003)
6. Lotter, T., Vary, P.: Speech Enhancement by MAP Spectral Amplitude Estimation Using a Super-Gaussian SpeechModel. EURASIP Journal on Applied Signal Processing 7, 1110–1126 (2005)
7. Ephraim, Y., Cohen, I.: Recent advancements in Speech Enhancement. In: The Electrical Engineering Handbook. CRC Press, Boca Raton (2005)
8. Erkelens, J., Hendriks, R., Heusdens, R., Jensen, J.: Minimum Mean-Square Error Estimation of Discrete Fourier Coefficients With Generalized Gamma Priors. IEEE Trans ASSP 15, 1741–1752 (2007)
9. Erkelens, J.S.: Speech Enhancement based on Rayleigh mixture modeling of Speech Spectral Amplitude Distributions. In: Proc. EUSIPCO 2007, pp. 65–69 (2007)
10. Hirsch, G.: Aurora-5 Experimental Framework for the Performance Evaluation of Speech Recognition in Case of a Hands-free Speech Input in Noisy Environments (2007)
11. Astudillo, R.F., Kolossa, D., Orglmeister, R.: Accounting for the Uncertainty of Speech Estimates in the Complex Domain for Minimum Mean Square Error Speech Enhancement. In: Proc. Interspeech, pp. 2491–2494 (2009)

Improving Keyword Spotting with a Tandem BLSTM-DBN Architecture

Martin Wöllmer[1], Florian Eyben[1], Alex Graves[2], Björn Schuller[1], and Gerhard Rigoll[1]

[1] Institute for Human-Machine Communication, Technische Universität München, Germany
{woellmer,eyben,schuller,rigoll}@tum.de
[2] Institute for Computer Science VI, Technische Universität München, Germany

Abstract. We propose a novel architecture for keyword spotting which is composed of a Dynamic Bayesian Network (DBN) and a bidirectional Long Short-Term Memory (BLSTM) recurrent neural net. The DBN uses a hidden garbage variable as well as the concept of switching parents to discriminate between keywords and arbitrary speech. Contextual information is incorporated by a BLSTM network, providing a discrete phoneme prediction feature for the DBN. Together with continuous acoustic features, the discrete BLSTM output is processed by the DBN which detects keywords. Due to the flexible design of our Tandem BLSTM-DBN recognizer, new keywords can be added to the vocabulary without having to re-train the model. Further, our concept does not require the training of an explicit garbage model. Experiments on the TIMIT corpus show that incorporating a BLSTM network into the DBN architecture can increase true positive rates by up to 10 %.

Keywords: Keyword Spotting, Long Short-Term Memory, Dynamic Bayesian Networks.

1 Introduction

Keyword spotting aims at detecting one or more predefined keywords in a given speech utterance. In recent years keyword spotting has found many applications, e.g. in voice command detectors, information retrieval systems, or embodied conversational agents. Hidden Markov Model (HMM) based keyword spotting systems [9] usually require keyword HMMs and a *garbage* HMM to model both, keywords and non-keyword parts of the speech sequence. However, the design of the garbage HMM is a non-trivial task. Using whole word models for keyword and garbage HMMs presumes that there are enough occurrences of the keywords in the training corpus and suffers from low flexibility since new keywords cannot be added to the system without having to re-train it. Modeling phonemes instead of whole words offers the possibility to design a garbage HMM that connects all phoneme models but implies that the garbage HMM can potentially model any phoneme sequence, including the keyword itself.

J. Solé-Casals and V. Zaiats (Eds.): NOLISP 2009, LNAI 5933, pp. 68–75, 2010.
© Springer-Verlag Berlin Heidelberg 2010

In this paper we present a new Dynamic Bayesian Network (DBN) design which can be used for robust keyword spotting and overcomes most of the drawbacks of other approaches. Dynamic Bayesian Networks offer a flexible statistical framework that is increasingly applied for speech recognition tasks [2,1] since it allows for conceptual deviations from the conventional HMM architecture. Our keyword spotter does not need a trained garbage model and is robust with respect to phoneme recognition errors. Unlike large vocabulary speech recognition systems, our technique does not require a language model but only the keyword phonemizations. Thereby we use a hidden garbage variable and the concept of *switching parents* [1] to model either a keyword or arbitrary speech.

In order to integrate contextual information into the keyword spotter, we extend our DBN architecture to a Tandem recognizer that uses the phoneme predictions of a bidirectional Long Short-Term Memory (BLSTM) recurrent neural net together with conventional MFCC features. Tandem architectures which combine the output of a discriminatively trained neural net with dynamic classifiers such as HMMs have been successfully used for speech recognition tasks and are getting more and more popular [6,8]. BLSTM networks efficiently exploit past and future context and have been proven to outperform standard methods of modeling contextual information such as triphone HMMs [4]. As shown in [12], the framewise phoneme predictions of a BLSTM network can enhance the performance of a discriminative keyword spotter. In [3] a BLSTM based keyword spotter trained on a fixed set of keywords is introduced. However, this approach requires re-training of the net as soon as new keywords are added to the vocabulary, and gets increasingly complex if the keyword vocabulary grows. The keyword spotting architecture proposed herein can be seen as an extension of the graphical model for spoken term detection we introduced in [13]. Thus, we aim at combining the flexibility of our DBN architecture with the ability of a BLSTM network to capture long-range time dependencies and the advantages of Tandem speech modeling.

The structure of this paper is as follows: Section 2 reviews the principle of DBNs and BLSTMs as the two main components of our keyword spotter. Section 3 explains the architecture of our Tandem recognizer while experimental results are presented in Section 4. Concluding remarks are mentioned in Section 5.

2   Keyword Spotter Components

Our Tandem keyword spotter architecture consists of two major components: a Dynamic Bayesian Network processing observed speech feature vectors to discriminate between keywords and non-keyword speech, and a BLSTM network which takes in to account contextual information to provide an additional discrete feature for the DBN. The following sections will shortly review the basic principle of DBNs and BLSTMs.

2.1 Dynamic Bayesian Network

Dynamic Bayesian Networks can be interpreted as graphical models $G(V, E)$ which consist of a set of nodes V and edges E. Nodes represent random variables which can be either hidden or observed. Edges - or rather *missing* edges - encode conditional independence assumptions that are used to determine valid factorizations of the joint probability distribution. Conventional Hidden Markov Model approaches can be interpreted as *implicit* graph representations using a single Markov chain together with an integer state to represent all contextual and control information determining the allowable sequencing. In this work however, we decided for the *explicit* approach [2], where information such as the current phoneme, the indication of a phoneme transition, or the position within a word is expressed by random variables.

2.2 Bidirectional LSTM Network

The basic idea of bidirectional recurrent neural networks [10] is to use two recurrent network layers, one that processes the training sequence forwards and one that processes it backwards. Both networks are connected to the same output layer, which therefore has access to complete information about the data points before and after the current point in the sequence. The amount of context information that the network actually uses is learned during training, and does not have to be specified beforehand.

Analysis of the error flow in conventional recurrent neural nets (RNNs) resulted in the finding that long time lags are inaccessible to existing RNNs since the backpropagated error either blows up or decays over time (vanishing gradient problem). This led to the introduction of Long Short Term Memory (LSTM) RNNs [7]. An LSTM layer is composed of recurrently connected memory blocks, each of which contains one or more recurrently connected memory cells, along with three multiplicative 'gate' units: the input, output, and forget gates. The gates perform functions analogous to read, write, and reset operations. More specifically, the cell input is multiplied by the activation of the input gate, the cell output by that of the output gate, and the previous cell values by the forget gate. Their effect is to allow the network to store and retrieve information over long periods of time.

Combining bidirectional networks with LSTM gives bidirectional LSTM, which has demonstrated excellent performance in phoneme recognition [4], keyword spotting [3], and emotion recognition [11]. A detailed explanation of BLSTM networks can be found in [5].

3 Architecture

The Tandem BLSTM-DBN architecture we used for keyword spotting is depicted in Figure 1. The network is composed of five different layers and hierarchy levels respectively: a word layer, a phoneme layer, a state layer, the observed features, and the BLSTM layer (nodes inside the grey shaded box). For the sake of simplicity only a simple LSTM layer, consisting of inputs i_t, a hidden layer h_t, and outputs o_t, is shown in Figure 1, instead of the more complex bidirectional LSTM which would contain two RNNs.

The following random variables are defined for every time step t: q_t denotes the phoneme identity, q_t^{ps} represents the position within the phoneme, q_t^{tr} indicates a phoneme transition, s_t is the current state with s_t^{tr} indicating a state transition, and x_t denotes the observed acoustic features. The variables w_t, w_t^{ps}, and w_t^{tr} are identity, position, and transition variables for the word layer of the DBN whereas a hidden *garbage variable* g_t indicates whether the current word is a keyword or not. A second observed variable b_t contains the phoneme prediction of the BLSTM. Figure 1 displays hidden variables as circles and observed variables as squares. Deterministic conditional probability functions (CPFs) are represented by straight lines and zig-zagged lines correspond to random CPFs. Dotted lines refer to so-called *switching parents* [1], which allow a variable's parents to change conditioned on the current value of the switching parent. Thereby a switching parent can not only change the set of parents but also the implementation (i.e. the CPF) of a parent. The bold dashed lines in the LSTM layer do not represent statistical relations but simple data streams.

Assuming a speech sequence of length T, the DBN structure specifies the factorization

$$p(g_{1:T}, w_{1:T}, w_{1:T}^{tr}, w_{1:T}^{ps}, q_{1:T}, q_{1:T}^{tr}, q_{1:T}^{ps}, s_{1:T}^{tr}, s_{1:T}, x_{1:T}, b_{1:T}) =$$

$$\prod_{t=1}^{T} p(x_t|s_t)p(b_t|s_t)f(s_t|q_t^{ps}, q_t)p(s_t^{tr}|s_t)f(q_t^{tr}|q_t^{ps}, q_t, s_t^{tr})f(w_t^{tr}|q_t^{tr}, w_t^{ps}, w_t)$$

$$f(g_t|w_t)f(q_1^{ps})p(q_1|w_1^{ps}, w_1, g_1)f(w_1^{ps})p(w_1)\prod_{t=2}^{T} f(q_t^{ps}|s_{t-1}^{tr}, q_{t-1}^{ps}, q_{t-1}^{tr})$$

$$p(w_t|w_{t-1}^{tr}, w_{t-1})p(q_t|q_{t-1}^{tr}, q_{t-1}, w_t^{ps}, w_t, g_t)f(w_t^{ps}|q_{t-1}^{tr}, w_{t-1}^{ps}, w_{t-1}^{tr}) \qquad (1)$$

with $p(\cdot)$ denoting random conditional probability functions and $f(\cdot)$ describing deterministic CPFs.

The size of the BLSTM input layer i_t corresponds to the dimensionality of the acoustic feature vector x_t whereas the vector o_t contains one probability score for each of the P different phonemes at each time step. b_t is the index of the most likely phoneme:

$$b_t = \max_{o_t}(o_{t,1}, ..., o_{t,j}, ..., o_{t,P}) \qquad (2)$$

The CPFs $p(x_t|s_t)$ are described by Gaussian mixtures as common in an HMM system. Together with $p(b_t|s_t)$ and $p(s_t^{tr}|s_t)$, they are learnt via EM training. Thereby s_t^{tr} is a binary variable, indicating whether a state transition takes place or not. Since the current state is known with certainty, given the phoneme and the phoneme position, $f(s_t|q_t^{ps}, q_t)$ is purely deterministic. A phoneme transition occurs whenever $s_t^{tr} = 1$ and $q_t^{ps} = S$ provided that S denotes the number of

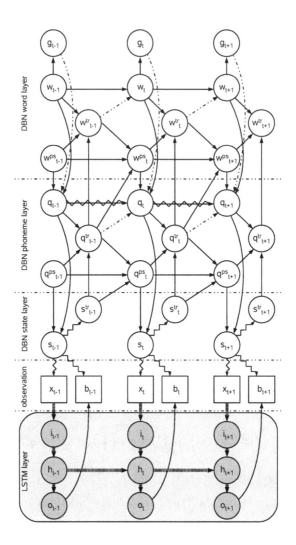

Fig. 1. Structure of the Tandem BLSTM-DBN keyword spotter

states of a phoneme. This is expressed by the function $f(q_t^{tr}|q_t^{ps}, q_t, s_t^{tr})$. The phoneme position q_t^{ps} is known with certainty if s_{t-1}^{tr}, q_{t-1}^{ps}, and q_{t-1}^{tr} are given.

The hidden variable w_t can take values in the range $w_t = 0...K$ with K being the number of different keywords in the vocabulary. In case $w_t = 0$ the model is in the *garbage state* which means that no keyword is uttered at that time. The variable g_t is then equal to one. w_{t-1}^{tr} is a switching parent of w_t: if no word transition is indicated, w_t is equal to w_{t-1}. Otherwise a word bigram specifies the CPF $p(w_t|w_{t-1}^{tr} = 1, w_{t-1})$. In our experiments we simplified the word bigram to a zerogram which makes each keyword equally likely. Yet, we introduced differing a priori likelihoods for keywords and garbage phonemes:

$$p(w_t = 1 : K | w^{tr}_{t-1} = 1) = \frac{K \cdot 10^a}{K \cdot 10^a + 1} \qquad (3)$$

and

$$p(w_t = 0 | w^{tr}_{t-1} = 1) = \frac{1}{K \cdot 10^a + 1}. \qquad (4)$$

The parameter a can be used to adjust the trade-off between true positives and false positives. Setting $a = 0$ means that the a priori probability of a keyword and the probability that the current phoneme does not belong to a keyword are equal. Adjusting $a > 0$ implies a more aggressive search for keywords, leading to higher true positive and false positive rates. The CPFs $f(w^{tr}_t | q^{tr}_t, w^{ps}_t, w_t)$ and $f(w^{ps}_t | q^{tr}_{t-1}, w^{ps}_{t-1}, w^{tr}_{t-1})$ are similar to the phoneme layer of the DBN (i.e. the CPFs for q^{tr}_t and q^{ps}_t). However, we assume that "garbage words" always consist of only one phoneme, meaning that if $g_t = 1$, a word transition occurs as soon as $q^{tr}_t = 1$. Consequently w^{ps}_t is always zero if the model is in the garbage state. The variable q_t has two switching parents: q^{tr}_{t-1} and g_t. Similar to the word layer, q_t is equal to q_{t-1} if $q^{tr}_{t-1} = 0$. Otherwise, the switching parent g_t determines the parents of q_t. In case $g_t = 0$ - meaning that the current word is a keyword - q_t is a deterministic function of the current keyword w_t and the position within the keyword w^{ps}_t. If the model is in the garbage state, q_t only depends on q_{t-1} in a way that phoneme transitions between identical phonemes are forbidden.

Note that the design of the CPF $p(q_t | q^{tr}_{t-1}, q_{t-1}, w^{ps}_t, w_t, g_t)$ entails that the DBN will strongly tend to choose $g_t = 0$ (i.e. it will detect a keyword) once a phoneme sequence that corresponds to a keyword is observed. Decoding such an observation while being in the garbage state $g_t = 1$ would lead to "phoneme transition penalties" since the CPF $p(q_t | q^{tr}_{t-1} = 1, q_{t-1}, w^{ps}_t, w_t, g_t = 1)$ contains probabilities less than one. By contrast, $p(q_t | q^{tr}_{t-1} = 1, w^{ps}_t, w_t, g_t = 0)$ is deterministic, introducing no likelihood penalties at phoneme borders.

4   Experiments

Our keyword spotter was trained and evaluated on the TIMIT corpus. The feature vectors consisted of cepstral mean normalized MFCC coefficients 1 to 12, energy, as well as first and second order delta coefficients. For the training of the BLSTM, 200 utterances of the TIMIT training split were used as validation set while the net was trained on the remaining training sequences. The BLSTM input layer had a size of 39 (one for each MFCC feature) and the size of the output layer was also 39 since we used the reduced set of 39 TIMIT phonemes. Both hidden LSTM layers contained 100 memory blocks of one cell each. To improve generalization, zero mean Gaussian noise with standard deviation 0.6 was added to the inputs during training. We used a learning rate of 10^{-5} and a momentum of 0.9.

The independently trained BLSTM network was then incorporated into the DBN in order to allow the training of the CPFs $p(b_t | s_t)$. During the first training cycle of the DBN, phonemes were trained framewisely using the training portion of the TIMIT corpus. Thereby all Gaussian mixtures were split once 0.02%

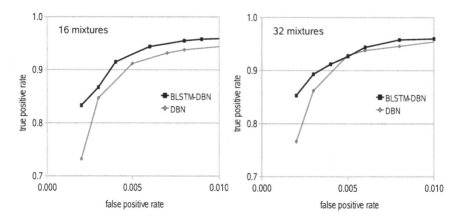

Fig. 2. Part of the ROC curve for the DBN keyword spotter and the Tandem BLSTM-DBN approach using different values for the trade-off parameter a. Left side: 16 Gaussian mixtures; right side 32 Gaussian mixtures.

convergence was reached until the number of mixtures per state increased to 16 and 32 respectively. In the second training cycle segmentation constraints were relaxed, whereas no further mixture splitting was conducted. Phoneme models were composed of three hidden states each.

We randomly chose 60 keywords from the TIMIT corpus to evaluate the keyword spotter. The used dictionary allowed for multiple pronunciations. The trade-off parameter a (see Equation 3) was varied between 0 and 10.

Figure 2 shows a part of the Receiver Operating Characteristics (ROC) curves for the DBN and the Tandem BLSTM-DBN keyword spotter, displaying the true positive rate (tpr) as a function of the false positive rate (fpr). Note that due to the design of the recognizer, the full ROC curve - ending at an operating point tpr=1 and fpr=1 - cannot be determined, since the model does not include a confidence threshold that can be set to an arbitrarily low value. The most significant performance gain of context modeling via BLSTM predictions occurs at an operating point with a false positive rate of 0.2%: there, the Tandem approach can increase the true positive rate by 10%. Conducting the McNemar's test revealed that the performance difference between the BLSTM-DBN and the DBN is statistically significant at a common significance level of 0.01. For higher values of the trade-off parameter a, implying a more aggressive search for keywords, the performance gap becomes smaller as more phoneme confusions are tolerated when seeking for keywords.

5   Conclusion

This paper introduced a Tandem BLSTM-DBN keyword spotter that makes use of the phoneme predictions generated by a bidirectional Long Short-Term Memory recurrent neural net. We showed that the incorporation of contextual information via BLSTM networks leads to significantly improved keyword spotting results.

Future works might include a combination of triphone and BLSTM modeling as well as processing the entire vector of BLSTM output activations instead of exclusively using the most likely phoneme prediction.

Acknowledgements. The research leading to these results has received funding from the European Community's Seventh Framework Programme (FP7/2007-2013) under grant agreement No. 211486 (SEMAINE).

References

1. Bilmes, J. A.: Graphical models and automatic speech recognition. In: Mathematical Foundations of Speech and Language Processing (2003)
2. Bilmes, J.A., Bartels, C.: Graphical model architectures for speech recognition. IEEE Signal Processing Magazine 22(5), 89–100 (2005)
3. Fernández, S., Graves, A., Schmidhuber, J.: An Application of Recurrent Neural Networks to Discriminative Keyword Spotting. In: de Sá, J.M., Alexandre, L.A., Duch, W., Mandic, D.P. (eds.) ICANN 2007. LNCS, vol. 4669, pp. 220–229. Springer, Heidelberg (2007)
4. Graves, A., Fernandez, S., Schmidhuber, J.: Bidirectional LSTM networks for improved phoneme classification and recognition. In: Proc. of ICANN, Warsaw, Poland, vol. 18(5-6), pp. 602–610 (2005)
5. Graves, A.: Supervised sequence labelling with recurrent neural networks. Phd thesis, Technische Universität München (2008)
6. Hermansky, H., Ellis, D.P.W., Sharma, S.: Tandem connectionist feature extraction for conventional HMM systems. In: Proc. of ICASSP, Istanbul, Turkey, vol. 3, pp. 1635–1638 (2000)
7. Hochreiter, S., Schmidhuber, J.: Long short-term memory. Neural Computation 9(8), 1735–1780 (1997)
8. Ketabdar, H., Bourlard, H.: Enhanced phone posteriors for improving speech recognition systems. IDIAP-RR, no. 39 (2008)
9. Rose, R.C., Paul, D.B.: A hidden markov model based keyword recognition system. In: Proc. of ICASSP, Albuquerque, NM, USA, vol. 1, pp. 129–132 (1990)
10. Schuster, M., Paliwal, K.K.: Bidirectional recurrent neural networks. IEEE Transactions on Signal Processing 45, 2673–2681 (1997)
11. Wöllmer, M., Eyben, F., Reiter, S., Schuller, B., Cox, C., Douglas-Cowie, E., Cowie, R.: Abandoning Emotion Classes - Towards Continuous Emotion Recognition with Modelling of Long-Range Dependencies. In: Proc. of Interspeech, Brisbane, Australia, pp. 597–600 (2008)
12. Wöllmer, M., Eyben, F., Keshet, J., Graves, A., Schuller, B., Rigoll, G.: Robust discriminative keyword spotting for emotionally colored spontaneous speech using bidirectional LSTM networks. In: Proc. of ICASSP, Taipei, Taiwan (2009)
13. Wöllmer, M., Eyben, F., Schuller, B., Rigoll, G.: Robust vocabulary independent keyword spotting with graphical models. In: Proc. of ASRU, Merano, Italy (2009)

Score Function for Voice Activity Detection

Jordi Solé-Casals, Pere Martí-Puig, Ramon Reig-Bolaño, and Vladimir Zaiats

Digital Technologies Group, University of Vic, Sagrada Família 7,
08500 Vic, Spain
`{jordi.sole,pere.marti,ramon.reig,vladimir.zaiats}@uvic.cat`

Abstract. In this paper we explore the use of non-linear transformations in order to improve the performance of an entropy based voice activity detector (VAD). The idea of using a non-linear transformation comes from some previous work done in speech linear prediction (LPC) field based in source separation techniques, where the score function was added into the classical equations in order to take into account the real distribution of the signal. We explore the possibility of estimating the entropy of frames after calculating its score function, instead of using original frames. We observe that if signal is clean, estimated entropy is essentially the same; but if signal is noisy transformed frames (with score function) are able to give different entropy if the frame is voiced against unvoiced ones. Experimental results show that this fact permits to detect voice activity under high noise, where simple entropy method fails.

Keywords: VAD, score function, entropy, speech.

1 Introduction

In speech and speaker recognition a fast and accurate detection of the speech signal in noise environment is important because the presence of non-voice segments or the omission of voice segments can degrade the recognition performance [1-3]. On the other hand in a noise environment there are a set of phonemes that are easily masked, and the problem of detecting the presence of voice cannot be solved easily. The problem is further complicated by the fact that the noise in the environment can be time variant and can have different spectral properties and energy variations. Also there are limitations on the admissible delay between the input signal, and the decision of the presence or absence of voice.

In the last several decades, a number of endpoint detection methods have been developed. According to [4] we can categorize approximately these methods into two classes. One is based on thresholds [4-7]. Generally, this kind of method first extracts the acoustic features for each frame of signals and then compares these values of features with preset thresholds to classify each frame. The other is pattern-matching method [8-9] that needs estimate the model parameters of speech and noise signal. The detection process is similar to a recognition process. Compared with pattern-matching method, thresholds-based method does not need keep much training data and train models and is simpler and faster.

J. Solé-Casals and V. Zaiats (Eds.): NOLISP 2009, LNAI 5933, pp. 76–83, 2010.

Endpoint detection by thresholds-based method is a typical classification problem. In order to achieve satisfied classification results, it is the most important to select appropriate features. Many experiments have proved that shortterm energy and zero-crossing rate fail under low SNR conditions. It is desirable to find other robust features superior to short-term energy and zero-crossing rate. J. L. Shen [10] first used the entropy that is broadly used in the field of coding theory on endpoint detection. Entropy is a metric of uncertainty for random variables, thus it is definite that the entropy of speeches is different from that of noise signals because of the inherent characteristics of speech spectrums.

However, it is found that the basic spectral entropy of speech varies to different degrees when the spectrum of speech is contaminated by different noise signals especially high noise signals. The varieties make it difficult to determine the thresholds. Moreover, the basic spectral entropy of various noises disturbs the detection process. It is expected that there exists a way by which it is possible that (1) the entropy of various noise signals approaches to one another under the same SNR condition, (2) the curve of noise entropy is flat, and (3) the entropy of speech signals differs from that of noise signals obviously.

This paper investigates different non-linear transforamtions on the input signal to improve voice activity detection based on spectral entropy. Preliminary experimental results shown that it is possible to improve basic spectral entropy, specially in the presence of non-gaussian noise or colored noise.

2 Entropy

Originally, the entropy was defined for information sources by Shannon [11] and is a measure of the uncertainty associated with a random variable. Is defined as:

$$H(S) = -\sum_{i=1}^{N} p(s_i) \log p(s_i) \tag{1}$$

where $S = [s_1, ..., s_i, ..., s_N]$ are all the possible values of the discrete random variable S, and $p(s_i)$ is the probability mass function of S.

In case of speech, for certain phonemes, the energy is concentrated in a few frequency bands, and therefore will have low entropy as the signal spectrum is more organized during speech segments; while in the case of noise with flat spectrum or low pass noise, the entropy will be higher. The measure of entropy is defined in the spectral energy domain as:

$$p_j(k) = \frac{|S_j(k)|}{\sum_{m=1}^{N} |S_j(m)|} \tag{2}$$

where $S_j(k)$ is the kth DFT coefficient in the jth frame. Then the measure of entropy is defined in the spectral energy domain as:

$$H(j) = -\sum_{i=1}^{N} p_j(k) \log p_j(k) \tag{3}$$

As $H(j)$ is maximum when S_j is a white noise and minimum (null) when it is a pure tone, the entropy of the noise frame is not dependent upon the noise level and the threshold can be estimated a priori. Under this observation, the entropy based method is well suited for speech detection in white or quasi-white noises, but will perform poorly for colored noises or non-Gaussian noises. We will see that applying some nonlinear function on the signal the entropy based method can deal with these cases.

3 Exploring Score Function as Non-linear Transformation

Inspired in BSS/ICA algorithms [see 12 and references therein] or blind linear/non-linear deconvolution [13-14], we propose to use score function to non-linearly modify the signal before calculating entropy for VAD process. What we expect is that as score function is related to pdf of the signal, we will enhance the difference between voice and non voice frames, even in very noisy environments.

3.1 Score Function

Given a vector Y, the so-called score is defined as:

$$\psi_Y(u) = \frac{\partial \log p_Y(u)}{\partial u} = \frac{p'_Y(u)}{p_Y(u)} \tag{4}$$

Since we are concerned by nonparametric estimation, we will use a kernel density estimator [15]. This estimator is easy to implement and has a very flexible form, but suffers from the difficulty of the choice of the kernel bandwidths. Formally we estimate $p_Y(u)$ by:

$$\hat{p}_Y(u) = \frac{1}{hT} \sum_{t=1}^{T} K\left(\frac{u - y(t)}{h}\right) \tag{5}$$

from which we get an estimate of $\psi_Y(u)$ by $\psi_Y(u) = \dfrac{\hat{p}'_Y(u)}{\hat{p}_Y(u)}$. Many kernel shapes can be good candidates, for our experiments we used the Gaussian kernel. A "*quick and dirty*" method for the choice of the bandwidth consists in using the rule of thumb $h = 1.06\hat{\sigma}T^{-\frac{1}{5}}$. Other estimators may be found, and used, but experimentally we noticed that the proposed estimator works fine.

3.2 Other Functions

In many BSS/ICA algorithms, score function is approximated by a fixed function, depending on the sub-Gaussian or super-Gaussian character of the signals. In this case, functions like $\tanh(u)$, $u\exp\left(-\frac{u^2}{2}\right)$ or u^3 are used instead of calculating the true score function $\psi_Y(u)$. As a preliminary analysis only results with the true score function will be presented in this paper, but these more simple functions can maybe be a good candidates, especially for real-time applications, as they avoid the task of estimate the true score function.

4 Proposed Method

The proposed method in order to explore these non-linear functions for VAD is shown in figure 1. The signal is framed and score function is estimated for each frame, using this output as the input to the next block (entropy calculation) instead of the original frame. What we are interested is looking at this entropy of the scored frame compared with the original frame (without score function).

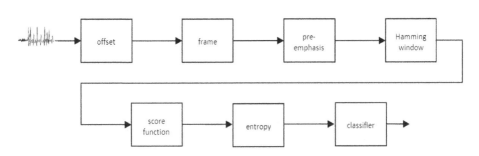

Fig. 1. Block diagram of the proposed method

Pre-processing stage will be done according to ETSI standard [16]. According to that, a notch filtering operation is applied to the digital samples of the input speech signal s_{in} to remove their DC offset, producing the offset-free input signal s_{of} :

$$s_{of}(n) = s_{in}(n) - s_{in}(n-1) + 0.999 s_{of}(n-1) \tag{6}$$

The signal is framed in a 25 ms frame length, that corresponds to 200 samples for a sampling rate $f_s = 8\,kHz$, with frame shift interval of 10 ms, that corresponds to 80 samples for a sampling rate $f_s = 8\,kHz$. A pre-emphasis filter is applied to the framed offset-free input signal,

$$s_{pe}(n) = s_{of}(n) - 0.97 s_{of}(n-1) \tag{7}$$

and finally a Hamming window is applied to the output of the pre-emphasis block. Once obtaining windowed frame of N samples, score function is estimated according to eq. 4 and 5, and then spectral entropy is computed by means of eq. 3. The final decision (voiced frame – unvoiced frame) is taken by means of a threshold, even if more complex and better rules can be considered, but here we are only exploring the differences between estimated entropy by using score function or not, in order to facilitate the classifier block.

5 Experiments

Several experiments are done in order to investigate the performance of the system. First of all, we are interested in looking how looks like a scored frame compared to a

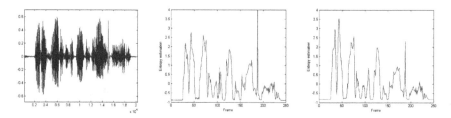

Fig. 2. Signal input (left) and estimated entropy (without score function on the middle, and with score function on the right)

simple frame. In figure 2 we can see a voiced signal, its estimated entropy and its estimated entropy over the score function. In this case, when the voice signal is clean (good SNR), we can observe a similar shape of the entropy for the original frames and scored frames.

If we add Gaussian noise to the signal, the results begin to be different, as we can see in figure 3, where we show the input signal (top left) and the clean signal for shake of clarity (down left), and the estimated entropy without and with score function. Even if noise is very high, we can observe that entropy is different in the parts of signal containing speech, but of course the difference is not as clear as in figure 2. Also we can observe that results without and with score function are not as similar as before. If noise is much harder, entropy estimation does not permit to distinguish between noise and speech, and then no voice activity can be detected.

On the other hand, if noise is uniform we can obtain better results for the estimation of entropy with score function. Results in figure 4 are obtained with uniform noise, and we observe that without score function we cannot distinguish between noise and speech, while it can be done with score function.

Using a simple threshold on the estimated entropy, we can make a decision on the signal, in order to decide if the frame is a voiced or unvoiced one. Of course, more elaborated procedures must be used instead of a simple trigger, as explained in the literature, but here for the sake of simplicity we will present some results only with a threshold.

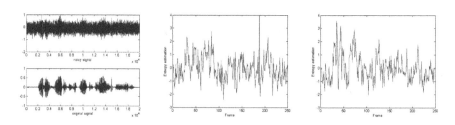

Fig. 3. Signal input (top left) with Gaussian noise, and estimated entropy (without score function on the middle, and with score function on the right)

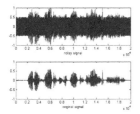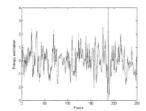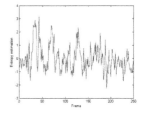

Fig. 4. Signal input (top left) with uniform noise, and estimated entropy (without score function on the middle, and with score function on the right)

Figure 5 shows the results obtained without score function (left) and with score function (right), with a clean speech signal (no noise added). We can see that asimple threshold can give us good results and that they are very similar, as the estimated entropy with and without score function are (approximately) equivalent, as we showed in figure 2.

On the other hand, if speech signal is noisy, and due to the fact that estimated entropy is no more equal without or with score function, the detection of speech is much more hard and different results are obtained using ore don't using score function. Results for this case are presented in figure 6.

In this case, scored frames give better results and hence the voice is better detected even if it is hidden by noise. Of course VAD doesn't gives perfect results, as we can see comparing the detection presented in fig. 6 with the true speech signal, plotted on the bottom of the figure for the sake of clarity, but this can be improved by designing a better classifier, as mentioned before.

6 Conclusions

In this paper, the use of non-linear transformations for improve a voice activity detector is explored.

Score function is used as non-linear transformation, estimated by means of a Gaussian kernel, and entropy is used as a criterion to decide if a frame is voiced or unvoiced.

If speech signal is clean, results are essentially the same, due to the fact that score function doesn't change the entropy of the signal. But in the case of noisy speech signal, the estimated entropy is no more equivalent, hence giving different results. Is in this case where frames pre-processed with score function give better results and voice can be detected into a very noisy signal.

Future work will be done exploring other non-linear transformation, trying to simplify and reduce the complexity of the system in order to be implemented in real-time applications. On the other hand, classifier will also be improved deriving some heuristic rules, for example, or by using more complex systems as neural networks, in order to minimise incorrect activity detections.

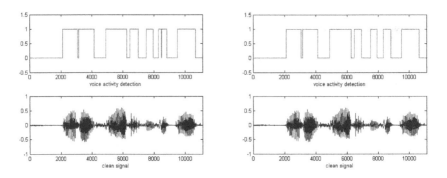

Fig. 5. Voice activity detection obtained with a simple threshold. On the left, estimating the entropy without score function. On the right, estimating the entropy with score function. As the estimated entropy is essentially the same, results are very coincident.

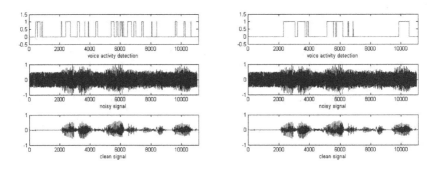

Fig. 6. Voice activity detection obtained with a simple threshold. On the left, estimating the entropy without score function. On the right, estimating the entropy with score function. As now the speech signal is noisy, the estimated entropy is different and hence the detection is also different. We can observe that scored signal gives better results.

Acknowledgments. This work has been supported by the University of Vic under the grant R0904.

References

1. Ying, G.S., Mitchell, C.D., Jamieson, L.H.: Endpoint Detection of Isolated Utterances Based on a Modified Teager Energy Measurement. In: Proc. ICASSP II, pp. 732–735 (1993)
2. Shen, J.-l., Hung, J.-w., Lee, L.-s.: Robust Entropy-based Endpoint Detection for Speech Recognition in Noisy Environments. In: Proc. ICSLP CD-ROM 1998 (1998)
3. Shin, W.-H., Lee, B.-S., Lee, Y.-K., Lee, J.-S.: Speech/Non-Speech Classification Using Multiple Features For Robust Endpoint Detection. In: Proc. ICASSP, pp. 1399–1402 (2000)

4. Jia, C., Xu, B.: An improved Entropy-based endpoint detection algorithm. In: Proc. ICSLP (2002)
5. Shin, W.-H., Lee, B.-S., Lee, Y.-K., Lee, J.-S.: Speech/non-speech classification using multiple features for robust endpoint detection. In: Proc. ICASSP (2000)
6. Van Gerven, S., Xie, F.: A Comparative study of speech detection methods. In: European Conference on Speech, Communication and Techonlogy (1997)
7. Hariharan, R., Häkkinen, J., Laurila, K.: Robust end-of-utterance detection for real-time speech recognition applications. In: Proc. ICASSP (2001)
8. Acero, A., Crespo, C., De la Torre, C., Torrecilla, J.: Robust HMM-based endpoint detector. In: Proc. ICASSP (1994)
9. Kosmides, E., Dermatas, E., Kokkinakis, G.: Stochastic endpoint detection in noisy speech. In: SPECOM Workshop, pp. 109–114 (1997)
10. Shen, J., Hung, J., Lee, L.: Robust entropybased endpoint detection for speech recognition in noisy environments. In: Proc. ICSLP, Sydney (1998)
11. Shannon, C.E.: A mathematical theory of communication. Bell System Technical Journal 27, 379–423, 623–656 (1948)
12. Hyvärinen, A., Karhunen, J., Oja, E.: Independent Component Analysis. John Wiley & Sons, Chichester (2001)
13. Solé-Casals, J., Taleb, A., Jutten, C.: Parametric Approach to Blind Deconvolution of Nonlinear Channels. Neurocomputing 48, 339–355 (2002)
14. Solé-Casals, J., Monte, E., Taleb, A., Jutten, C.: Source separation techniques applied to speech linear prediction. In: Proc. ICSLP (2000)
15. Härdle, W.: Smoothing Techniques with implementation in S. Springer, Heidelberg (1990)
16. ETSI standard doc., ETSI ES 201 108 V1.1.3 (2003-2009)

Digital Watermarking: New Speech and Image Applications

Marcos Faundez-Zanuy

Escola Universitària Politècnica de Mataró (Adscrita a la UPC)
08303 MATARO (BARCELONA), Spain
faundez@eupmt.es
http://www.eupmt.es/veu

Abstract. Digital watermarking is a powerful way to protect author's copyrights and to prove that a digital content has not been modified. Watermarking is the process of altering the original data file, allowing for the subsequent recovery of embedded auxiliary data called watermark. A subscriber, with knowledge of the watermark and how it is recovered, can determine (to a certain extent) whether significant changes have occurred within the data file. Researchers have also developed techniques that can help to identify portions of an image or speech signal that may contain unauthorized changes. This can be useful, for instance, in front of a court. In this paper, we present some recent applications in image and speech processing. We think that some more applications will appear in the future, and specially, commercial devices able to insert a digital watermark in a multimedia content. This is the milestone to popularize these systems.

1 Introduction

A watermark is always some additional information connected to a host signal. When the watermark is invisible for the unauthorized person, it is used for data hiding.

There are several, and often conflicting characteristics which are of interest for watermarking [1]:

- Quantity: A user typically wants to embed as much data as possible in a given media content. One important factor for the embedding capacity is the bandwidth of the host signal, i.e. there is much more additional data that can be put into a video stream, than on a telephone speech channel.
- Robustness: The hidden data needs to be invariant when a "host" signal is subject to a certain amount of distortion and transformation, e.g., lossy compression, channel noise, filtering, resampling, cropping, encoding, digital-to-analog (D/A) conversion, and analog-to-digital (A/D) conversion, etc.
- Security: The embedded data should be immune to intentional modification attempts to remove or manipulate the embedded data. For authentication purposes this would be of high importance.
- Transparency: The perceptual impact of invisible watermarks on the host content should be minimized. The host signal should not be objectionably degraded and the embedded data should be minimally perceptible.

J. Solé-Casals and V. Zaiats (Eds.): NOLISP 2009, LNAI 5933, pp. 84–89, 2010.

- Recovery: The embedded data should be self-clocking or arbitrarily reentrant. This ensures that the embedded data can be recovered when only fragments of the host signal are available, e.g., if a sound bite is extracted from an interview, data embedded in the audio segment can be recovered. This feature also facilitates automatic decoding of the hidden data, since there is no need to refer to the original host signal.

As mentioned before, some of these properties are competitive. For example, one cannot maximize security and minimize data quantity insertion. Depending on the application, for some characteristics a certain trade off has to be made in favor of more important properties.

There has been a lot of research effort in image watermarking and to a lesser extent in music. The main commercial application to date is copyright protection. For instance, commercial products such as Corel Draw, Photoshop and Paint Shop Pro include an option for "watermarking" based on Digimarc technology. Figure 1 shows a snapshot of this menu.

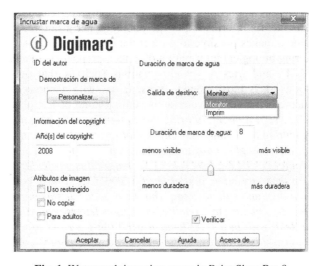

Fig. 1. Watermark insertion menu in Paint Shop Pro 8

The paper is organized as follows: section two describes a couple of applications for image signals. Section three is devoted to speech signals applications and section four summarizes the main conclusions and future work.

2   Image Processing Applications

In addition to Digimark technology described in the previous section, there is a couple of interesting applications. The first one is devoted to video surveillance [2]. In a critical video surveillance system, the authenticity guarantee of the images is essential. Thanks to the digital processing of the sources, TRedess has developed a very important complement for the video surveillance systems. With a patented software in

Fig. 2. Original (left), manipulated (center), and detection result. The inserted but on the right of the image is clearly detected thanks to the watermark information.

the EU, USA and Japan, the system is able to record images with a watermark that allow, in a later analysis, the identification of the altered parts of a recording with a very high reliability. Figure 2 shows an example of an original image (on the left), a manipulated image by inserting a bus not present in the original scene (on the centre), and the result of the detection (on the right).

Another very recent application is a digital photograph camera developed by Canon that inserts the iris information of the photographer inside the photo [3]. This reflex digital camera makes possible to protect the copyright of photographic images by reliably acquiring biological information of a photographer (US Patent Application No. 2008/0025574), and tries to avoid situations such as that described in [4]: _rebekka, one of Flickr's most popular users, is out with a photo of a screenshot where some other user named vulcanacar is selling some of her photos on iStockphoto. From _rebekka: "*I thought i'd bring this to light as a warning and wake up call to folks. I'm not putting the blame with iStockphoto per se, but still, this is a problem that is becoming increasingly annoying, for everyone that uses the internet to show-case work. I mean for crying out loud, out of 31 images this particular user has on his "portfolio", 25 of them are mine, and at least 3 are of me. I have contacted a copyright lawyer about this, and will be doing what I can to tackle this problem in the best way possible, (please, nobody start sending any angry letters or anything of that sort, this needs to be dealt with in a level-headed way) but I wanted to tell people about this as well, because this is always happening, and your photos could be on there illegally as well. You never know.*"

3   Speech Processing Applications

Although speech watermarking has not been so developed as image applications, we have performed several applications in the past. In [5-6] we proposed a speech watermarking system and we studied, respectively, the implications on speaker verification and identification system [7]. This proposal was based on a formed design for air traffic control, in order to secure the communications between an airplane and the controller at the airport [8].

In [9] we have recently developed an application that tries to solve a similar problem to that described in figure 2. We insert a watermark for forensic examination of speech recordings. This watermark consists of the time stamp (36 bits per second), which is the clock information shown in table 1. Although one second could be

Table 1. Time encoding in 36 bit (year, month, day, hour, minute, second)

	Year	Month	Day	Hour		
Format	y	m	d	h	m	s
Range	[2000,3023]	[1,12]	[1,31]	[0,23]	[0,59]	[0,59]
bits	10	4	5	5	6	6

enough to change some relevant words (for instance a name), such replacement could be detected: in order to detect the correct time stamp the whole second is necessary. If someone removes a portion of this second, the inserted portion will not contain the correct watermark information and the modification will be detected due to the incorrect time stamp recovery. Even though we do not exactly know what has been changed, we do know that something has been change within that second.

Although we have described the system with 36 bit (for pedagogical purposes), 30 bits are enough to encode the time stamp information. For instance, with the next MATLAB code:

```
y2k = 730485;  % (offset, days from year 0)
time = (fix((rem(datenum(clock)-y2k,1)*86400)+1)); % current time in seconds
date = fix(datenum(clock)-y2k); % current date in days from y2k
ds  =str2num([dec2bin(time,17),dec2bin(date,13)]')';
          % concatenate time and date and calculate binary
DATETIME =
round(datevec(bin2dec(num2str(ds(18:end),-6))+y2k+bin2dec(num2str(ds(1:17),-
6))./86400)); % decode date/time binary
    disp('File watermarking time:')
disp([num2str(DATETIME(3),'%02d'),'.',num2str(DATETIME(2),'%02d'),'.',num2
str(DATETIME(1)),'
',num2str(DATETIME(4),'%02d'),':',num2str(DATETIME(5),'%02d'),':',num2str(D
ATETIME(6),'%02d')])
```

Table 2 presents an example of an original untampered speech fragment, starting at 10:23:20 on June the 25th 2009, and several possible modifications (insertion, deletion and modification). As can be seen, these modifications are detected due to incorrect time stamp value, which is underlined. Only when the file integrity is assured, the watermark can be detected in a correlative way.

The actual setting could not be enough for some applications, but it is not a problem to increase the resolution by means of the following ways (one of them or both of them simultaneously):

   a)   If we decide to remove the year information a significant reduction is achieved and less than 30 bits are necessary.

   b)   If we increase the watermark power (decreasing the signal to watermark ratio), more bits can be inserted, and the whole time stamp can be detected in less than one second.

Table 2. Different scenarios for the speech watermarking system. Only the untouched speech file is able to reproduce the correct stream of time-stamps. Any modification, i.e. deletion, insertion, replacement results in a random or mission timestamp is detected.

Scenario	Stream of Time-stamps (y-m-d ; h:m:s)			
Original and untampered	2009-06-25 10:23:20	2009-06-25 10:23:21	2009-06-25 10:23:22	2009-06-25 10:23:23
One second in missing (in sync with the watermark)	2009-06-25 10:23:20	2009-06-25 10:23:22	2009-06-25 10:23:23	2009-06-25 10:23:24
One second in missing (at arbitrary position)	2009-06-25 10:23:20	2012-12-06 11:41:52	2009-06-25 10:23:23	2009-06-25 10:23:24
Part of a second is missing	2009-06-25 10:23:20	2029-03-03 03:56:32	2009-06-25 10:23:22	2009-06-25 10:23:23
Part of a second is replaced	2009-06-25 10:23:20	2001-2-31 22:43:11	2009-06-25 10:23:22	2009-06-25 10:23:23
More than a second is inserted	2009-06-25 10:23:20	2001-12-30 22:43:11	2013-08-18 09:45:23	2009-06-25 10:23:22

4 Conclusions

In this paper we have presented several watermarking applications for image and speech signals. Commercial applications can be bought for images. Although there are no commercial systems for speech, we think that in a near future these applications will be available thanks to the experimental systems developed in laboratory conditions.

Acknowledgement

This work has been supported by the Spanish project MEC TEC2006-13141-C03/TCM and COST-2102.

References

1. Bender, W., et al.: Techniques for data hiding. IBM systems journal 35(3-4), 313–336 (1996)
2. http://www.tredess.com/en/catalogo_int.php?id_cat=2&id=17

3. http://www.photographybay.com/2008/02/09/
   canon-iris-registration-watermark/
4. http://thomashawk.com/2008/02/
   istockphoto-user-vulcanacar-photo-thief.html
5. Faundez-Zanuy, M., Hagmüller, M., Kubin, G.: Speaker verification security improvement by means of speech watermarking. Speech Commun. 48(12), 1608–1619 (2006)
6. Faundez-Zanuy, M., Hagmüller, M., Kubin, G.: Speaker identification security improvement by means of speech watermarking. Pattern Recognition 40(11), 3027–3034 (2007)
7. Faundez-Zanuy, M., Monte-Moreno, E.: State-of-the-art in speaker recognition. IEEE Aerospace and Electronic Systems Magazine 20(5), 7–12 (2005)
8. Hering, H., Hagmüller, M., Kubin, G.: Safety and security increase for air traffic management through unnoticeable watermark aircraft identification tag transmitted with the VHF voice communication. In: Proc. 22nd Digital Avionics Systems Conference (DASC), Indianapolis, Indiana, October 2003, pp. 4_E_2_1–10 (2003)
9. Faundez-Zanuy, M., Lucena-Molina, J.J., Hagmueller, M.: Speech watermarking: an approach for the forensic analysis of digital telephonic recordings. Journal of Forensic Sciences (in press)

Advances in Ataxia SCA-2 Diagnosis Using Independent Component Analysis

Rodolfo V. García[1], Fernando Rojas[2], Carlos G. Puntonet[2], Belén San Román[3], Luís Velázquez[4], and Roberto Rodríguez[4]

[1] Network Department, University of Holguín. Cuba. Spanish MAEC-AECID fellowship
[2] Department of Computer Architecture and Technology, University of Granada. Spain
frojas@atc.ugr.es
[3] PhD Student. University of Granada. Spain
[4] Centre for the Research and Rehabilitation of Hereditary Ataxias "Carlos J. Finlay", Holguín. Cuba

Abstract. This work discusses a new approach for ataxia SCA-2 diagnosis based in the application of independent component analysis to the data obtained by electro-oculography in several experiments carried out over healthy and sick subjects. Abnormalities in the oculomotor system are well known clinical symptoms in patients of several neurodegenerative diseases, including modifications in latency, peak velocity, and deviation in saccadic movements, causing changes in the waveform of the patient response. The changes in the morphology waveform suggest a higher degree of statistic independence in sick patients when compared to healthy individuals regarding the patient response to the visual saccadic stimulus modeled by means of digital generated saccade waveforms. The electro-oculogram records of six patients diagnosed with ataxia SCA2 (a neurodegenerative hereditary disease) and six healthy subjects used as control were processed to extract saccades.

1 Ataxia SCA2 Incidence

The ocular movement records have been widely used in processing and classification of biological signals and pathological conditions: clinical sleep scoring [1], cerebellar dysfunctions [2, 3], diagnosis of the visual system [4, 5], amongst others, also in human computer interface and visual guided devices [6-8]. The Spino Cerebellar Ataxia type 2 (SCA-2) is an autosomal dominant cerebellar hereditary ataxia with the highest prevalence in Cuba, reporting up to 43 cases per 100,000 inhabitants in the province of Holguin. In most families there is clinical and neuropathological evidence of additional involvement of brainstem, basal ganglia, spinal cord, and the peripheral nervous system [9]. This form of ataxia occurs commonly in persons of Spanish ancestry in north-eastern Cuba, a figure much higher than that found in western Cuba or in other parts of the world. The high prevalence is probably the result of a founder effect, but might be due to an interaction between a mutant gene and an unidentified environmental neurotoxin [10, 11].

Several studies have reported oculomotor abnormalities in SCA2 [9, 11, 12]. Specifically, slowness of saccades has been suggested as a relatively characteristic

J. Solé-Casals and V. Zaiats (Eds.): NOLISP 2009, LNAI 5933, pp. 90–94, 2010.
© Springer-Verlag Berlin Heidelberg 2010

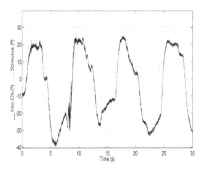 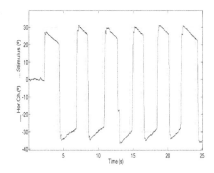

Fig. 1. Original records of horizontal saccade movements at 60°, in a patient of severe ataxia (left) and a healthy subject (right)

finding in SCA2[11, 13]. This fact determines significant differences in saccade morphology between healthy individuals and patients with SCA-2, mainly for 60° of stimulus amplitude. The electro-oculographical records are quite different in healthy individuals and patients with a severe ataxia as it is shown in Figure 1 for a pursuit experiment (saccadic eye movement).

2   Independent Component Analysis Applied to EOG Data

Independent component analysis is aimed to find a linear transformation given by a matrix $\mathbf{W}$, so that the random variables $\mathbf{y}_i$, $(i=1,...,n)$ of $\mathbf{y}=[\mathbf{y}_1,...,\mathbf{y}_n]$ are as independent as possible in:

$$\mathbf{y}(t) = \mathbf{W} \cdot \mathbf{x}(t) \tag{1}$$

This linear blind source separation approach is suitable for the signals obtained by the EOG, as well as in other medical analysis such as electroencephalography (EEG), electrocardiography (ECG), magneto-encephalography (MEG), and functional magnetic resonance imaging (fMRI) [12, 14].

In the analysis of EOG oriented to the detection of SCA2 experts anticipate two possible behaviors of the individuals: sick and healthy conduct. During an experiment over a healthy subject, the horizontal movement of the eye is expected to follow the stimulus signal. Therefore, the horizontal eye movement and the stimulus will hold a direct dependence between them, i.e. the signals are not independent. In contrast, a sick individual may present a more chaotic response, depending on the severity of the disease. Consequently, the subject response will not depend in such a high degree on the stimulus signal, and the signals are independent (or at least, "not so dependent").

Therefore, the proposed methodology uses independent component analysis as a classification algorithm criterion: if the independence measure (normally mutual information) reveals independence between the individual response and the stimulus signal, then it is rather possible that the individual presents some degree of ataxia or related disease.

The proposed algorithm for ataxia SCA-2 diagnosis will go along the following steps:

1. Set both horizontal response and stimulus signal in the same phase, i.e. correct the delay between the stimulus change and the saccade.
2. Normalize signals (x).
3. Apply ICA algorithm. Any well known ICA algorithm may be applied at this point (FastICA [15], Jade [16], GaBSS [17], etc.).
4. Normalize estimations (y)
5. Calculate error measure between estimations (y) and mixtures (x) according to the root mean square error expression:

$$RMSE(\mathbf{x}_i, \mathbf{y}_i) = \sqrt{\frac{\sum_{t=0}^{N} [x(t) - y(t)]^2}{N}} \tag{2}$$

6. Depending on the obtained error measure, a simple categorization algorithm (such as C-means) may be applied in order to classify individuals. Otherwise, a human expert may help in subject categorization based on the ICA results.

3   Experimental Results and Discussion

The electro-oculogram recordings of six patients with severe ataxia and six healthy subjects diagnosed and classified in the "Centre for the Research and Rehabilitation of Hereditary Ataxias (CIRAH)" were used in order to perform the analysis of repeated ocular saccadic movement tests for 10°, 20°, 30° and 60° divergence stimuli.

All the records were carried out by the medical staff of CIRAH. Each individual was placed in a chair, with a head fixation device to avoid head movements, the variables were collected by a two channel electronystagmograph (Otoscreen, Jaeger-Toennies). Recording conditions were set as follows: electrodes of silver chloride placed in the external borders of right eye (active electrode) and left eye (reference electrode), high pass filtering 0.002 Hz, low pass filtering 20 Hz, sensitivity 200 µV/division, and sampling frequency 200 Hz. For stimulus generation a black screen

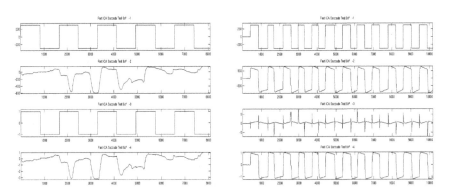

Fig. 2. Stimulus (1), response (2) and ICA components (3 and 4) obtained at 60° of stimulation for patients (left) and control subjects (right)

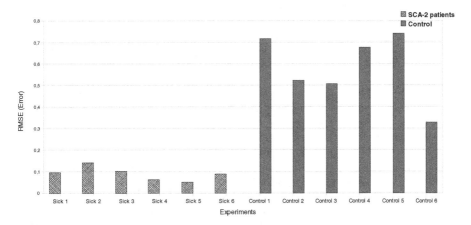

Fig. 3. Root mean squared error between the estimations and the sources after the application of the algorithm to EOG data corresponding to SCA-2 patients (left) and control subjects (right)

CRT display showing a white circular target with an angular size of 0.7° was used. The stimulus and patient response data are automatically stored in ASCII files by Otoscreen electronystagmograph.

The patient response was filtered using a median filter, to obtain a clean waveform of the patient response, afterwards it was phased with the stimulus. Finally FastICA was applied to get the independent components (See Figure 2).

As Figure 3 depicts, results show that the error measure obtained for SCA-2 patients is clearly differentiable for the same measure obtained for control subjects. That is due to the fact mentioned in the hypothesis that if the independence measure reveals independence between the individual response and the stimulus signal, then it is possible that the individual presents some degree of ataxia. When the original signals (stimulus and response) were independent, the estimations are close to those sources and, therefore, the RMS error decreases.

The results were obtained from six control subjects and six patients. Confirming our hypothesis, starting from electro-oculography experiments, patients showed a different behavior in terms of the visual response to a fixed stimulus (see Figure 2 and Figure 3). Therefore, after applying our proposed approach to the raw EOG data, classification and diagnosis can be made easily by simple human inspection of the results. Nevertheless, further research in this line may help in the categorization of the several stages of severity of SCA-2.

4 Conclusion

The proposed method starts from the assumption that the response to a visual stimulus is different in a healthy individual when compared to the response of an individual afflicted by SCA-2. In the later situation, the response from the individual is not dependent on the visual stimulus, so that the ICA algorithm estimations will be similar to the obtained observations. This criterion has shown to be suitable in order to distinguish between sick (patients) and healthy (control) individuals.

References

1. Virkkala, J., Hasan, J., Värri, A., Himanen, S.-L., Müller, K.: Automatic sleep stage classification using two-channel electro-oculography. Journal of Neuroscience Methods 166, 109–115 (2007)
2. Spicker, S., Schulz Jr., B., Petersen, D., Fetter, M., Klockgether, T., Dichgans, J.: Fixation instability and oculomotor abnormalities in Friedreich's ataxia. Journal of Neurology 242, 517–521 (1995)
3. Yokota, T., Hayashi, H., Hirose, K., Tanabe, H.: Unusual blink reflex with four components in a patient with periodic ataxia. Journal of Neurology 237, 313–315 (1990)
4. Güven, A.l., Kara, S.: Classification of electro-oculogram signals using artificial neural network. Expert Systems with Applications 31, 199–205 (2006)
5. Irving, E.L., Steinbach, M.J., Lillakas, L., Babu, R.J., Hutchings, N.: Horizontal Saccade Dynamics across the Human Life Span. Invest. Ophthalmol. Vis. Sci. 47, 2478–2484 (2006)
6. Kumar, D., Poole, E.: Classification of EOG for human computer interface. In: Annual International Conference of the IEEE Engineering in Medicine and Biology - Proceedings, vol. 1, pp. 64–67 (2002)
7. Brunner, S., Hanke, S., Wassertheuer, S., Hochgatterer, A.: EOG Pattern Recognition Trial for a Human Computer Interface. Universal Access in Human-Computer Interaction. Ambient Interaction, 769–776 (2007)
8. Komogortsev, O.V., Khan, J.I.: Eye movement prediction by Kalman filter with integrated linear horizontal oculomotor plant mechanical model. In: Proceedings of the 2008 symposium on Eye tracking research & applications, pp. 229–236. ACM, New York (2008)
9. Bürk, K., Fetter, M., Abele, M., Laccone, F., Brice, A., Dichgans, J., Klockgether, T.: Autosomal dominant cerebellar ataxia type I: oculomotor abnormalities in families with SCA1, SCA2, and SCA3. Journal of Neurology 246, 789–797 (1999)
10. Orozco, G., Estrada, R., Perry, T.L., Araña, J., Fernandez, R., Gonzalez-Quevedo, A., Galarraga, J., Hansen, S.: Dominantly inherited olivopontocerebellar atrophy from eastern Cuba: Clinical, neuropathological, and biochemical findings. Journal of the Neurological Sciences 93, 37–50 (1989)
11. Velázquez, L.: Ataxia espino cerebelosa tipo 2. In: Principales aspectos neurofisiológicos en el diagnóstico, pronóstico y evaluación de la enfermedad. Ediciones Holguín, Holguín (2006)
12. Vaya, C., Rieta, J.J., Sanchez, C., Moratal, D.: Convolutive Blind Source Separation Algorithms Applied to the Electrocardiogram of Atrial Fibrillation: Study of Performance. IEEE Transactions on Biomedical Engineering 54, 1530–1533 (2007)
13. Klostermann, W., Zühlke, C., Heide, W., Kömpf, D., Wessel, K.: Slow saccades and other eye movement disorders in spinocerebellar atrophy type 1. Journal of Neurology 244, 105–111 (1997)
14. Anemüller, J.r., Sejnowski, T.J., Makeig, S.: Complex independent component analysis of frequency-domain electroencephalographic data. Neural Networks 16, 1311–1323 (2003)
15. Hyvarinen, A., Oja, E.: A Fast Fixed-Point Algorithm for Independent Component Analysis. Neural Computation 9, 1483–1492 (1997)
16. Cardoso, J.-F.: Source separation using higher order moments. In: Proceedings of ICASSP, IEEE International Conference on Acoustics, Speech and Signal Processing, vol. 4, pp. 2109–2112 (1989)
17. Rojas, F., Puntonet, C.G., Rodríguez-Álvarez, M., Rojas, I., Martín-Clemente, R.: Blind source separation in post-nonlinear mixtures using competitive learning, simulated annealing, and a genetic algorithm. IEEE Transactions on Systems, Man and Cybernetics Part C: Applications and Reviews 34, 407–416 (2004)

Spectral Multi-scale Product Analysis for Pitch Estimation from Noisy Speech Signal

Mohamed Anouar Ben Messaoud, Aïcha Bouzid, and Noureddine Ellouze

Department of Electrical Engineering, National School of Engineers of Tunis,
Le Belvédère BP. 37, 1002 Tunis, Tunisia
anouar.benmessaoud@yahoo.fr, bouzidacha@yahoo.fr,
n.ellouze@enit.rnu.tn

Abstract. In this work, we present an algorithm for estimating the fundamental frequency in speech signals. Our approach is based on the spectral multi-scale product analysis. It consists of operating a short Fourier transform on the speech multi-scale product. The multi-scale product is based on making the product of the speech wavelet transform coefficients at three successive dyadic scales. The wavelet used is the quadratic spline function with a support of 0.8 ms. We estimate the pitch for each time frame based on its multi-scale product harmonic structure. We evaluate our approach on the Keele database. Experimental results show the effectiveness of our method presenting a good performance surpassing other algorithms. Besides, the proposed approach is robust for noisy speech.

Keywords: Speech, wavelet transform, multi-scale product, spectral analysis, fundamental frequency.

1 Introduction

One of the most salient features of human speech is its harmonic structure. During voiced speech segments, the regular glottal excitation of the vocal tract produces energy at the fundamental frequency F_0 and its multiples [1].

A reliable algorithm for pitch tracking is critical in many speech processing tasks such as prosody analysis, speech enhancement, and speech recognition.

Numerous pitch determination algorithms (PDAs) have been proposed [2] and are generally classified into three categories: time domain, frequency domain, and time frequency domain. Time domain PDAs directly examines the temporal structure of a signal wave form. Typically, peak and valley positions, zero crossings and autocorrelations are analysed for detecting the pitch period. Frequency domain PDAs distinguish the fundamental frequency by utilising the harmonic structure in the short-term analysis on band-filtered signals obtained via a multi-channel front-end [3].

Although many pitch determination algorithm (PDA) were proposed, there is still no reliable and useful algorithm which can be used sufficiently for many speech processing applications.

The difficulty of accurate and robust pitch estimation of speech is due to several reasons as the fast variation of the instantaneous pitch and formants.

J. Solé-Casals and V. Zaiats (Eds.): NOLISP 2009, LNAI 5933, pp. 95–102, 2010.

In this paper, we present and evaluate an algorithm for estimating the fundamental frequency in speech signals. Our proposed algorithm operates a spectral analysis of a derived speech signal. This analysis produces rays and one of these rays corresponds to the fundamental frequency. The derived speech signal is obtained by multiplying the wavelet transform coefficients of the speech signal at three successive dyadic scales.

The paper is presented as follows. After the introduction, we present the multi-scale product method used in this work to provide the derived speech signal. Section 3 introduces the spectral multi-scale product (SMP) approach for the fundamental frequency estimation. In section 4, we present evaluation results and compare them with results of existing algorithms for clean speech. Evaluation results are also presented for speech mixed with a white Gaussian noise at various SNR levels.

2 Multi-scale Product

The wavelet transform is a multi-scale analysis which has been shown to be very well suited for speech processing as glottal closure instant (GCI) detection, pitch estimation, speech enhancement and recognition and so on. Using the wavelet transform, a speech signal can be analyzed at specific scales corresponding to the range of human speech [4], [5], [6] and [7].

One of the most important wavelet transform applications is the signal singularity detection. In fact, continuous wavelet transforms produce modulus maxima at signal singularities allowing their localisation. However, one-scale analysis does not give a good precision. So, decision algorithm using multiple scales is proposed by different works to circumvent this problem [8] and [9].

The multi-scale product is essentially introduced to improve signal edge detection. A non linear combination of wavelet transform coefficients enhances edges and suppresses the spurious peaks.

This method was first used in image processing. This approach is based on the multiplication of wavelet transform coefficients at some scales. It attempts to enhance the peaks of the gradients caused by true edges, while suppressing false peaks caused by noise. Xu and al. rely on the variations in scale of the wavelet transform. They use multiplication of wavelet transform of the image at adjacent scales to distinguish important edges from noise [10]. Sadler and Swami [11] have studied multi-scale product method of signal in presence of noise. In wavelet domain, it is well known that edge structures are present at each sub-band, while noise decreases rapidly along the scales. It has been observed that multiplying the adjacent scales could sharpen edges while smoothing out noise [5] and [6]. In Bouzid and Ellouze, the Multi-scale Product (MP) is used to detect the glottal closure and opening instants from speech signal. The speech MP produces maxima at GOI and minima at GCI [9].

The multi-scale product [11] consists of making the product of wavelet transform coefficients of the function f(n) at some successive dyadic scales as follows

$$p(n) = \prod_{j=j_0}^{j=j_L} w_{2^j} f(n).$$ (1)

Where $w_{2^j} f(n)$ is the wavelet transform of the function f at scale 2^j.

Odd number of terms in p(n) preserves the sign of maxima. Choosing the product of three levels of wavelet decomposition is generally optimal and allows detection of small peaks.

In this work, we are motivated by the MP use because it provides a derived speech signal which is simpler to be analysed. The voiced speech multi-scale product has a

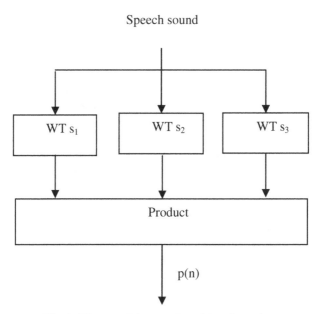

Fig. 1. Diagram of the speech multi-scale product

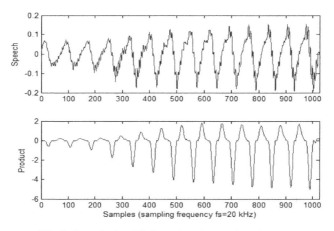

Fig. 2. Speech signal followed by its multi-scale product

periodic structure with more reinforced singularities marked by extrema. It has a structure that reminds the derivative laryngograph signal. So, a spectral analysis can be applied on the obtained signal.

The figure 2 depicts the speech signal followed by its MP. The MP reveals unambiguous maxima corresponding to GOI and clear minima corresponding to GCI.

3 Spectral Multi-scale Product for Pitch Estimation

We propose a new technique to estimate the fundamental frequency. The method is based on the spectral analysis of the speech multi-scale product. It can be decomposed into three essential steps, as shown in figure 3. The first step consists of computing the product of the speech sound wavelet transform coefficients. The wavelet used in this multi-scale product analysis is the quadratic spline function at scales $s_1 = 2^{-1}$, $s_2 = 2^0$ and $s_3 = 2^1$. The second step consists of calculating the fast Fourier transform (FFT) of the obtained signal over windows with a specific length of 4096 samples. In deed, the product is decomposed into frames of 1024 samples with an overlapping of 512 points at a sampling frequency of 20 kHz. The last step consists of looking for the maxima constituting the spectrum and the frequencies corresponding to these maxima. These maxima need to be classified to make a decision about the fundamental frequency value.

The first step is detailed in the previous section.

For the second step, the product p[n] is divided into frames of N length by multiplication with a sliding analysis window w[n]:

$$p_w[n, i] = p[n] w[n - i\Delta n] \qquad (2)$$

Where i is the window index, and Δn the overlap. The weighting w[n] is assumed to be non zero in the interval [0, N-1]. The frame length value N is chosen in such a way that, on the one hand, the parameters to be measured remain constant and, on the other hand, that there are enough samples of p[n] within the frame to guarantee reliable frequency parameter determination.

The choice of the windowing function influences the values of the short term parameters, the shorter the window the greater is its influence [12].

Each weighted block $p_w[n]$ is transformed in the spectral domain using Discrete Fourier Transform (FFT), in order to extract the spectral parameters $P_w[k]$, where k represents the index of the frequency ([0, N-1]). The DFT of each frame is expressed as follows:

$$P_w^i[k] = \sum_{n=0}^{N-1} p_w[n, i] e^{-j\frac{2\pi}{N} nk} \qquad (3)$$

Figure 4 illustrates the spectrum amplitude of the speech MP. The obtained signal shows spectral rays. In this case, the first ray has the greatest amplitude and its corresponding frequency matches with the signal fundamental frequency.

To ensure an accurate F0 estimation, we propose a strategy to select the correct spectral ray maxima that allows the fundamental frequency estimation and not the harmonics.

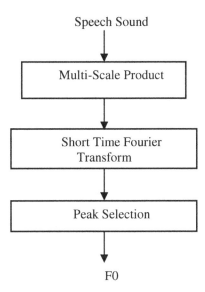

Speech Sound

Multi-Scale Product

Short Time Fourier
Transform

Peak Selection

F0

Fig. 3. Block diagram of the proposed approach for the fundamental frequency estimation

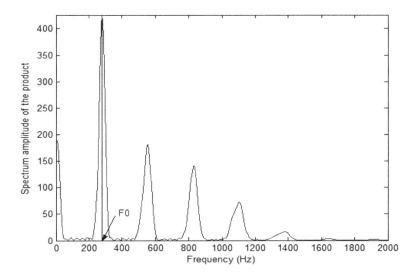

Fig. 4. Spectral multi-scale product of the speech signal

The third step consists of peak selection for the fundamental frequency estimation. All the maxima are localised and ranked for each window, the greatest maxima is selected with its harmonics, to constitute a first series. The second maximum is considered with its harmonics and the second series is formed. This is done for all maxima. The selected series has the greatest number of harmonics. The first peak gives the fundamental frequency.

Some pitch errors can be avoided by considering the continuity of the pitch between frame analyses, which improves the algorithm performance.

4   Evaluation

To evaluate the performance of our algorithm, we used the Keele pitch reference database [13]. This database consists of speech signals of five male and five female English speakers each reading the same phonetically balanced text with varying duration between about 30 and 40 seconds. The reference pitch estimation is based on a simultaneously recorded signal of a laryngograph. Uncertain frames are labelled using a negative flag. The authors of the database suggest ignoring these frames in performance comparisons. We use common performance measures for comparing pitch estimation algorithms: The gross pitch error (GPE) denotes the percentage of frames at which the estimation and the reference pitch differ by more than 20%. The root mean square error (RMS) is computed as the RMS difference in Hertz between the reference pitch and the estimation for all frames that are not GPE.

Table 1 presents evaluation results of the proposed algorithm (SMP) for clean speech. For comparison, we list results of other state of the art algorithms [1], [14], [15], [16] and [17] that are based on the same reference database.

Table 1. Evaluation results of the SMP algorithm and others for clean speech

Methods	GPE (%)	RMS (Hz)
SMP	*0.75*	*2.41*
NMF-HMM-PI [1]	1.06	3.7
NMF [14]	0.9	4.3
MLS [15]	1.5	4.68
Yin [16]	2.35	3.62
RAPT [17]	2.2	4.4

As can be seen, our approach yields good results encouraging us to use it in a noisy environment. In fact, the SMP method shows a low GPE rate and an interesting RMS value.

To test the robustness of the algorithm, we add a white Gaussian noise at different signal-to-noise ratios (SNRs) to the original signals. As depicted in tables 2 and 3, when the SNR level decreases, the SMP algorithm remains reliable even at -5dB and more efficient than the Yin algorithm. Besides, the SMP method presents the lowest RMS values showing its convenience for pitch estimation in hard situations.

Table 2. GPE of the proposed method SMP and the Yin algorithm for noisy speech

SNR levels	SMP	Yin
	GPE (%)	
5 dB	*1*	3.9
0dB	*1.2*	5.1
-5dB	*1.4*	5.9

Table 3. Fundamental frequency deviation RMS of the proposed method SMP and the Yin algorithm

SNR levels	SMP	Yin
	RMS (Hz)	
5 dB	*3.23*	4.1
0dB	*3.73*	5.7
-5Db	*4.67*	7.8

5 Conclusion

In this paper, we present a pitch estimation method that relies on the spectral analysis of the speech multi-scale product. The proposed approach can be summarised in three essential steps. First, we make the product of speech wavelet transform coefficients at three successive dyadic scales (The wavelet is the quadratic spline function with a support of 0.8 ms). Second, we compute the short time Fourier transform of the speech multi-scale product. Thirdly, we select the peak corresponding to the fundamental frequency satisfying some criteria. The experimental results show the robustness of our approach for noisy speech, and its efficiency for clean speech in comparison with the state-of-the-art algorithms.

Acknowledgments. The authors are very grateful to Alain de Cheveigné for providing the f0 estimation software (Yin algorithm).

References

1. Joho, D., Bennewitz, M., Behnke, S.: Pitch Estimation Using Models of Voiced Speech on Three Levels. In: 4th IEEE International Conference on Acoustics, Speech and Signal Processing ICASSP 2007, pp. 1077–1080. IEEE Press, Honolulu (2007)
2. Hess, W.J.: Pitch Determination of Speech Signals, pp. 373–383. Springer, Heidelberg (1983)
3. Wu, M., Wang, D., Brown, G.J.: A Multipitch Tracking Algorithm for Noisy Speech. IEEE Trans. Speech Audio Process. 11, 229–241 (2003)
4. Burrus, C.S., Gopinath, R.A., Guo, H.: Introduction to Wavelets and Wavelet Transform: A Primer. Prentice Hall, Englewood Cliffs (1998)
5. Mallat, S.: A Wavelet Tour of Signal Processing, 2nd edn. Academic Press, London (1999)
6. Berman, Z., Baras, J.S.: Properties of the Multiscale Maxima and Zero-crossings Representations. IEEE Trans. Signal Process. 41, 3216–3231 (1993)
7. Kadambe, S., Boudreaux-Bartels, G.F.: Application of the Wavelet Transform for Pitch Detection of Speech Signals. IEEE Trans. Information Theory 38, 917–924 (1992)
8. Bouzid, A., Ellouze, N.: Electroglottographic Measures Based on GCI and GOI Detection Using Multiscale Product. International Journal of Computers, Communications and Control 3, 21–32 (2008)
9. Bouzid, A., Ellouze, N.: Open Quotient Measurements Based on Multiscale Product of Speech Signal Wavelet Transform. Research Letters in Signal Processing, 1687–1691 (2007)

10. Xu, Y., Weaver, J.B., Healy, D.M., Lu, J.: Wavelet Transform Domain Filters: a Spatially Selective Noise Filtration Technique. IEEE Trans. Image Process. 3, 747–758 (1994)
11. Sadler, B.M., Swami, A.: Analysis of Multi-scale Products for Step Detection and Estimation. IEEE Trans. Information Theory 45, 1043–1051 (1999)
12. Shimamura, T., Takagi, H.: Noise-Robust Fundamental Frequency Extraction Method Based on Exponentiated Band-Limited Amplitude Spectrum. In: 47th IEEE International Midwest Symposium on Circuits and Systems MWSCAS 2004, pp. 141–144. IEEE Press, Hiroshima (2004)
13. Meyer, G., Plante, F., Ainsworth, W.A.: A Pitch Extraction Reference Database. In: 4th European Conference on Speech Communication and Technology EUROSPEECH 1995, Madrid, pp. 837–840 (1995)
14. Sha, F., Saul, L.K.: Real Time Pitch Determination of One or More Voices by Nonnegative Matrix Factorization. In: Saul, L.K., Weiss, Y., Bottou, L. (eds.) Advances in Neural Information Processing Systems, vol. 17, pp. 1233–1240. MIT Press, Cambridge (2005)
15. Sha, F., Burgoyne, J.A., Saul, L.K.: Multiband Statistical Learning for F0 Estimation in Speech. In: Proc. of the International Conference on Acoustics, Speech and Signal Processing ICASSP 2004, Montreal, pp. 661–664 (2004)
16. De Cheveigné, A., Kawahara, H.: YIN, a Fundamental Frequency Estimator for Speech and Music. J. Acoust. Soc. Am. 111, 1917–1930 (2002)
17. Talkin, D.: A Robust Algorithm for Pitch Tracking. In: Kleijn, W.B., Paliwal, K.K. (eds.) Speech Coding and Synthesis, pp. 495–518. Elsevier, Amsterdam (1995)

Automatic Formant Tracking Method Using Fourier Ridges

Imen Jemaa[1,2], Kaïs Ouni[1], and Yves Laprie[2]

[1] Unité de Recherche Traitement du Signal, Traitement de l'Image et Reconnaissance de Forme (99/UR/1119)
Ecole Nationale d'Ingénieurs de Tunis, BP.37, Le Belvédère 1002, Tunis, Tunisie
imen_jemaa@yahoo.fr, kais.ouni@enit.rnu.tn
[2] Equipe Parole, LORIA – BP 239 – 54506 Vandœuvre -lès- nancy, France
Yves.Laprie@loria.fr

Abstract. This paper develops a formant tracking method based on Fourier ridges detection. This work aims to improve the performance of formant tracking algorithm. In this method we have introduced a continuity constraint based on the computation of centre of gravity for a set of frequency formant candidates which leads to connect a frame of speech to its neighbours and thus to improve the robustness of track. The formant trajectories obtained by the algorithm proposed are compared to those of the hand edited formant Arabic database and those given by Praat with LPC data.

Keywords: Formant tracking, Fourier ridges, Speech processing, Centre of gravity, Formant labelling, Arabic database.

1 Introduction

A lot of research has focused on the development and improvement of algorithms for estimating and formant tracking. Indeed, robust formant tracks are used to identify vowels [1] and other vocalic sounds [2], to pilot formant synthesizers and, in some cases, to provide a speech recognition with additional data. Although, automatic formant tracking has a wide range of applications, it is still an open problem in speech analysis. Numerous works have been dedicated to develop algorithms for estimating formant frequencies from the acoustic signal. Most existing traditional methods are based on dynamic programming algorithms [3]. Formant tracking algorithms based on probabilistic approaches, particularly the Hidden Markov Models (HMM) [1], have been proposed in the same way. There are other methods like Linear Predictive Coding (LPC) spectra [4], these methods depend on the accuracy of the spectral peaks found by LPC.

Finally, other methods have received much attention in the research community in recent years, such as the technique of concurrent curves [5].

In this paper we describe our new automatic formant tracking algorithm based on the detection of Fourier ridges which are the maxima of spectrogram. This algorithm uses a continuity constraint by calculating the centre of gravity for a set of frequency

J. Solé-Casals and V. Zaiats (Eds.): NOLISP 2009, LNAI 5933, pp. 103–110, 2010.
© Springer-Verlag Berlin Heidelberg 2010

formant candidates. Then, to report a quantitative evaluation of the proposed algorithm to other automatic formant tracking methods; we developed a formant Arabic database as reference. Formant trajectories have been labeled by hand and checked by phonetician experts. This database is well balanced with respect to gender and phonetic contexts.

This paper is presented as follows, we present in section 2, the proposed formant tracking algorithm, in section 3, the description of the formant Arabic database labelling, in section 4 the results obtained by comparing the proposed Fourier ridges algorithm with other automatic formant tracking methods. Finally, we give some perspectives in section 5.

2   Formant Tracking Algorithm

The block diagram presented in Figure 1 describes the main steps of the proposed algorithm. Each element of the block diagram is briefly described below.

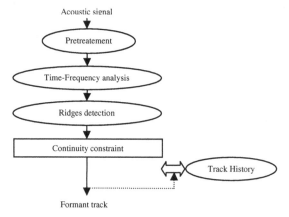

Fig. 1. Block diagram of the proposed formant tracking algorithm

2.1   Pretreatment

The sampled frequency used on speech signals in the database is FS=16 kHz. Since we are interested in the first three formants, we re-sampled the speech signal at Fs=8 kHz in order not to take into account formant candidates above 4 kHz and to optimize the computing time. Then to accentuate the high frequencies, a first order pre-emphasis filter of the form $1 - 0.98\,z^{-1}$ is applied on the speech signal.

2.2   Time-Frequency Analysis

The time-frequency analysis of the signal is carried by the module to the square of the windowed Fourier transform to obtain the spectrogram of the signal. The spectrogram used here is a wideband spectrogram witch smoothes the spectral envelope of the signal to show the temporal evolution of formants.

2.3   Ridges Detection

Since the formant frequencies vary as a function of time, they are simulated to the instantaneous frequencies of the signal. The algorithm calculates so the all instantaneous frequencies of the input signal which are considered as local maxima of its spectrogram [6]. Indeed, we try firstly to show how the instantaneous frequency is validated like local maxima of the spectrogram to the input signal. We generated the family of time-frequency atoms of the windowed Fourier transform noted $g_{u,\xi}$ by time translations and frequency modulations of a real and symmetric window $g(t)$.

This atom has a center frequency ξ and is symmetric with respect to u (translation factor). The windowed Fourier transform $Sf(u,\xi)$ is carried by the correlation of this atom with the input signal f.

It was shown that the instantaneous frequency $\phi'(t)$ of f , which it is the positive derived phase of this signal, is related to the windowed Fourier transform $Sf(u,\xi)$ if $\xi \geq 0$ (Eq.1).

$$Sf(u,\xi) = \frac{\sqrt{s}}{2} a(u) \exp^{i(\phi(u) - \xi(u))} \times \left[\hat{g}\left(s\left[\xi - \phi'(u)\right]\right) + \varepsilon(u,\xi) \right] \qquad (1)$$

Where s is a scaling which has been applied to the window Fourier g, $\hat{g}$ is the Fourier Transform, FT, of g, $a(t)$ is the analytic amplitude of f and $\varepsilon(u,\xi)$ is the corrective term. Since $|\hat{g}(\omega)|$ is maximum at $\omega = 0$, the Equation (1) shows that for each u, the spectrogram $|Sf(u,\xi)|^2$ is maximum at its center frequency $\xi(u) = \phi'(u)$. As a result, the instantaneous frequency is well validated like local maxima of the spectrogram. The windowed Fourier ridges are the maxima of the spectrogram at the FT points $(u,\xi(u))$. In each temporal window analysis, the algorithm detects thus all local maxima of the FT representation assimilated at Fourier ridges in the $(u,\xi(u))$ plane. The analysis window used here is Hamming window. The window duration is 6 ms and we consider an overlap equal to 50 %. We suppose, in this work, that the track concerns only the three first formant tracks, so in each temporal window analysis, we have split the set of frequency formant candidates detected previously into three wide bands respective for each formant track. The algorithm proceeds, then, to a parabolic interpolation and a threshold to remove ridges corresponding to small amplitude, because there may be artefacts from for example, "shadows" of other frequencies produced by the side-lobes of the window Fourier transform analysis or instantaneous frequency specific to the fundamental frequency F0 [7]. Then, we calculate the frequencies corresponding to the remaining ridge points. These frequencies are the formant candidates that might be chosen from to from the formant tracks.

2.4   Continuity Constraint

It is considered that in general, formants vary slowly as a function of time what leads to impose a continuity constraint in the process of selecting formant frequencies from the set of candidates. For the other algorithms based on LPC spectra, the continuity constraint used for each formant trajectory is the moving average of the LPC roots over its respective frequency band [8] [9]. In this algorithm we propose the calculation of the centre of gravity for a set of frequency formant candidates, detected by the ridge detection stage, as a continuity constraint between signal frames. Since the detection of ridges gives several candidates close together for one formant, the idea is to calculate the centre of gravity of the set of candidates located in the frequency band associated to the formant considered. The resulting frequency f is given by

$$\bar{f} = \frac{\sum_{i=1}^{n} p_i f_i}{\sum_{i=1}^{n} p_i} \tag{2}$$

Where f_i is the frequency of the ith candidate and p_i its spectral energy.

Considering centres of gravity instead of isolated candidates allows smooth formant transitions to be recovered.

3 The Formant Arabic Database Labelling

Given the interest of studies dedicated to formant tracking applied to the Arabic language and the lack of public formant Arabic database, we have decided to record and label in terms of formants a corpus of Standard Arabic pronounced by Tunisian speakers which will be publically available.

3.1 The Database Description

To build our corpus, we used a list of phonetically balanced Arabic sentences proposed by Boodraa and al. [10]. Most sentences of this database are extracted from Koran and Hadith. It consists of 20 lists of 10 short sentences. Each list of the corpus consists of 104 CV (C: consonant and V: vowel), i.e. 208 phonemes [10]. We recorded this corpus in a soundproof room. It comprises ten Tunisian speakers (five male speakers and five female speakers). The signal is digitized at a frequency of 16 kHz. This corpus contains 2000 sentences (200 different sentences pronounced by every speaker) either affirmative or interrogative. In this way, the database presents a well balanced selection of speakers, genders and phonemes. All the utterances of the corpus contain rich phonetic contexts and thus are a good collection of acoustic-phonetic phenomena that exhibit interesting formant trajectories [11]. In order to get accurate formant annotations, we have to verify every frequency value of formants with respect to phonemes uttered. Thus, to prepare our database, we phonetically annotated all the sentences of the corpus by hand using Winsnoori software [8]. A screenshot of this tool is shown in Fig.2. Once the annotation phase has been completed, the corpus has been reviewed by three phonetician experts to correct certain mistakes.

3.2 Formant Track Labelling

To facilitate the process of formant trajectory labelling in the database preparation, we first obtained a set of frequency formant candidates provided by the LPC roots [4] using Winsnoori [8]. Based on these initial estimative values, we drawn formant trajectories by hand by selecting relevant trajectories. Fig.2. shows an example of a sentence pronounced by a female speaker to illustrate the formant labelling process and results. We labeled the first three formants (F1, F2 and F3) every 4 ms and recorded them for each sentence of the database. The LPC order used is 18 and the temporal size of the window analysis is 4 ms to have a wideband spectrogram which shows the evolution of the formant trajectories better. More difficult situations arise when there are too many LPC candidates which are close to each other for two formant trajectories. But most of the difficulties arise for the frames where there is a lack of spectral prominences or when spectral prominences do not coincide with predicted resonances for consonantal segments. In this case, nominal consonant specific values are provided [12]. Finally, in order to verify the accuracy of the formant trajectories for each sentence, we have synthesized the sound with the corresponding three formant frequencies using Klatt synthesizer implemented in Winsnoori [8] to check whether the synthesized sentence matches the original one well. The evaluation was subjective though, since the authors were the only judges of the result quality.

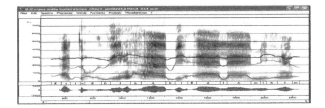

Fig. 2. Formant trajectories labeled (F1, F2 and F3) by hand and manually phonetic annotation for the sentence"أَتُؤُذِيهَا بِآلَامِهِمْ؟" "3atu3Di:ha: bi3a:la:mihim" (Would you harm her feelings with their pains) pronounced by a female speaker

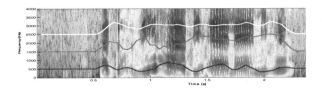

Fig. 3. Automatic formant trajectories (F1, F2 and F3) obtained by the Fourier ridge algorithm for the sentence "3atu3Di:ha: bi3a:la:mihim"(Would you harm her feelings with their pains)

4 Results and Discussion

We used our formant database as a reference to verify the accuracy of the proposed algorithm. To enable a visual comparison, Figures 2 and 3 show the automatic formant (F1, F2 and F3) tracking results for the same example sentence"أَتُؤُذِيهَا بِآلَامِهِمْ؟ " pronounced by a female speaker. The first figure is the reference obtained by hand, and the other result is obtained by our algorithm. It can be seen that for most of the vocalic portions in the utterances where the "dark/high energy" bands in the wideband spectrograms are clearly identifiable, the proposed method gives accurate results. To quantitatively evaluate the proposed algorithm, we compare it with the LPC method of Praat [9] using our formant database as a reference. The default LPC order used in Praat is 16. The size of the window analysis is 25 ms. The evaluation of each method consists in calculating the averaging absolute difference in (Hz) (Eq.3) and the standard deviation normalized with respect to manual reference values in (%) (Eq.4) for every formant trajectory (F1, F2 and F3). We thus examined results obtained for the short vowel /a/ within the syllable CV. Table 1 shows the results obtained on the vowel /a/ preceded by one consonant from every phonetic class for each formant trajectory (F1, F2 et F3) and for both tracking method. The CV occurrences were taken from the two following sentences pronounced by one male speaker: " وَاليًا عَرَفَ وَقَائِدَ" ("ɛarafa wa:liyan wa qa:3idan" which means "He knew a governor and a commander") and "هِيَ هُنَا لَقَدْ آبَتْ"("hiya huna: laqad 3a:bat" which means "She is here and she was pious").

$$D \, iff \ = \ \frac{1}{N} \times \sum_{p=1}^{N} \ \left| F_r (p) - F_c (p) \right| \ H z \tag{3}$$

$$\sigma \ = \ \sqrt{ \frac{1}{N} \sum_{p=1}^{N} \left(\frac{ \left| F_r (p) - F_c (p) \right| }{ F_r } \right)^2 } \tag{4}$$

Where F_r is the reference frequency, F_c the calculated frequency corresponding to both formant tracking methods and N the total number of points of each formant trajectory. The comparison of the errors shown in Table 1 shows that, globally, there is no big difference in terms of errors between the two automatic formant tracking methods except in some cases when the vowel /a/ is preceded 1) by a semi-vowel for F2 and F3, 2) by a tap for F3, 3) by a voiced plosive for F1 and F2. In these cases the algorithm proposed presents results close to the reference, and better results than the LPC method especially for F1 and F2 (except for F2 when the vowel /a/ is preceded by a voiceless plosive).

Table 1. Formant tracking errors measured for the short vowel /a/ within the syllable CV with different types of consonant

Formant Tracks		LPC F1	LPC F2	LPC F3	Fourier Ridges F1	Fourier Ridges F2	Fourier Ridges F3
Plosive voiced :	Diff	119	93	212	78	29	285
/d/	σ	42	37	65	18	6	71
Plosive voiceless :	Diff	71	58	67	75	110	17
/q/	σ	16	11	14	17	24	3
Fricative voiced /H/	Diff	49	48	36	28	49	33
	σ	8	7	6	4	8	6
Fricative voiceless : /f/	Diff	44	31	79	16	40	36
	σ	14	8	24	4	11	9
Nasal : /n/	Diff	97	30	101	73	47	166
	σ	22	7	25	15	13	46
Lateral : /l/	Diff	76	34	45	35	76	105
	σ	24	8	8	8	15	24
Tap : /r/	Diff	57	68	246	34	78	166
	σ	13	14		8	12	
Semi-vowel : /w/	Diff	98	91	140	92	32	89
	σ	31	49	52	18	7	20

We also notice that in most cases, the Fourier ridge algorithm presents large difference errors to the reference for F3. Tables 2 and 3 present formant tracking errors measured by averaging absolute difference and standard deviation between the reference and estimated values for each type of vowel, i.e. short vowels (/a/,/i/ and /u/) and long vowels (/A/, /I/ and /U/) pronounced respectively by two male speakers and two female speakers. Tests have been performed on four sentences: "عَرَفَ , " أَثُوْذِيهَا بِآلامِهِمْ؟", "هِيَ هُنَا لَقَدْ آبَتْ", "وَالِيَّا وَقائِدَ", cited above, and a last one أَسَرُونَا بِمُنْعَطَف ("3asaru:na: bimunɛatafin" which means "They captured us at a bend").

Table 2. Formant tracking errors measured for each type of vowel pronounced by two different male speakers

		MS1						MS2				
		LPC			Fourier Ridges			LPC			Fourier Ridges		
		F1	F2	F3	F1	F2	F3	F1	F2	F3	F1	F2	F3
a	Diff	44	31	79	16	40	36	34	44	152	17	18	103
	σ	14	8	24	4	11	8	9	12	49	5	4	25
A	Diff	52	91	89	38	71	82	58	114	63	79	82	34
	σ	14	45	29	9	14	18	13	23	14	16	18	7
i	Diff	35	49	58	28	25	91	28	53	161	26	190	192
	σ	12	18	20	9	9	28	12	20	73	10	64	69
I	Diff	42	67	64	170	198	12	64	47	83	45	150	166
	σ	13	19	21	46	54	4	20	14	25	15	41	48
u	Diff	57	82	89	118	95	99	64	55	90	91	80	28
	σ	15	20	28	32	23	41	10	20	31	31	26	9
U	Diff	65	83	265	133	62	160	86	191	215	58	93	187
	σ	19	22	64	41	17	52	40	125	103	19	29	86

Table 3. Formant tracking errors measured for each type of vowel pronounced by two different female speakers

		WS1						WS2				
		LPC			Fourier Ridges			LPC			Fourier Ridges		
		F1	F2	F3	F1	F2	F3	F1	F2	F3	F1	F2	F3
a	Diff	46	59	49	25	70	30	47	107	48	30	100	34
	σ	10	18	11	6	15	6	9	22	9	6	19	6
A	Diff	93	115	100	168	91	115	44	51	113	189	166	88
	σ	14	21	14	20	15	17	6	7	14	24	27	11
i	Diff	33	87	75	49	82	56	39	57	104	41	127	64
	σ	11	26	21	15	26	14	12	29	39	14	49	20
I	Diff	32	110	185	24	411	268	38	445	289	15	624	225
	σ	11	41	66	7	106	75	10	137	78	4	162	53
u	Diff	48	138	346	109	51	290	112	343	690	85	296	158
	σ	17	82	124	37	16	80	30	129	150	19	78	55
U	Diff	36	96	182	46	77	240	43	90	158	47	167	92
	σ	10	29	58	15	23	68	13	24	46	13	47	26

Table 2 shows that results are good especially for the vowels (/a/,/A/ et /i/) contrary to the (/I/,/u/ and /U/) probably because of their lower energy. We also notice that the proposed algorithm presents better results for the male speaker1 (MS1) than for (MS2). Table 3 shows that results are good for the vowels (/a/, /i/) for both tracking methods, especially for the female speaker 1 (WS1) contrary to (WS2) where there are some errors for F3 and F2 for both algorithms. However, results are not very good for the other vowels uttered by female speakers whatever the tracking method.

5 Conclusion

In this paper, we presented a new automatic formant tracking method based on detection of Fourier ridges. The constraints of continuity between speech frames are introduced by calculating of gravity centre of a combination of formant frequencies candidates. Furthermore, we report an exploratory use of an Arabic database to quantitatively evaluate the proposed algorithm. It provides accurate formant trajectories for F1 and F2 and results not as good for F3 in some cases.

Our future work will target the improvement of this method especially for high frequency formants.

Acknowledgement

This work is supported by the CMCU: Comité Mixte Franco-Tunisien de Coopération Universitaire (Research Project CMCU, 07G Code 1112).

References

1. Acero, A.: Formant Analysis and Synthesis using Hidden Markov Models. In: Proc. of the Eurospeech Conference, Budapest (1999)
2. Abdellaty Ali, A.M., Van der Spiegel, J., Mueller, P.: Robust Auditory-based Processing using the Average Localized Synchrony Detection. IEEE Transaction Speech and Audio Processing (2002)
3. Xia, K., Wilson, C.E.: A new Strategy of Formant Tracking based on Dynamic Programming. In: International Conference on Spoken Language Processing, ICSLP, Beijing, Chine (2000)
4. McCandless, S.: An algorithm for Automatic Formant extraction using Linear Prediction Spectra. Proc. IEEE 22, 135–141 (1974)
5. Laprie, Y.: A Concurrent Curve Strategy for Formant Tracking. In: Proc. of ICSLP, Jegu, Korea (2004)
6. Mallat, S.: A Wavelet Tour of Signal Processing. Academic Press, London (1999)
7. Châari, S., Ouni, K., Ellouze, N.: Wavelet Ridge Track Interpretation in Terms of Formants. In: International Conference on Speech and Language Processing, INTERSPEECH-ICSLP, Pittsburgh, Pennsylvania, USA, pp. 1017–1020 (2006)
8. http://www.loria.fr/~laprie/WinSnoori/
9. http://www.praat.org/
10. Boudraa, M., Boudraa, B., Guerin, B.: Twenty Lists of Ten Arabic Sentences for Assessment. In: Proc. of ACUSTICA, vol. 86, pp. 870–882 (2000)
11. Deng, L.: A Database of Vocal Tract Resonance Trajectories for Research in Speech Processing. In: Proc. of ICASSP (2006)
12. Ghazeli, S.: Back Consonants and Backing Coarticulation in Arabic. PhD, University of Texas, Austin (1977)

Robust Features for Speaker-Independent Speech Recognition Based on a Certain Class of Translation-Invariant Transformations*

Florian Müller and Alfred Mertins

Institute for Signal Processing, University of Lübeck, 23538 Lübeck, Germany
{mueller,mertins}@isip.uni-luebeck.de

Abstract. The spectral effects of vocal tract length (VTL) differences are one reason for the lower recognition rate of today's speaker-independent automatic speech recognition (ASR) systems compared to speaker-dependent ones. By using certain types of filter banks the VTL-related effects can be described by a translation in subband-index space. In this paper, nonlinear translation-invariant transformations that originally have been proposed in the field of pattern recognition are investigated for their applicability in speaker-independent ASR tasks. It is shown that the combination of different types of such transformations leads to features that are more robust against VTL changes than the standard mel-frequency cepstral coefficients and that they almost yield the performance of vocal tract length normalization without any adaption to individual speakers.

Keywords: Speech recognition, speaker-independency, translation-invariance.

1 Introduction

The vocal tract length (VTL) is a source of variability which causes the error rate of today's speaker-independent automatic speech recognition (ASR) systems to be two to three times higher than for speaker-dependent ASR systems [1]. Besides its shape it is the length of the vocal tract that determines the location of the resonance frequencies, commonly known as "formants". The formants determine the overall envelope of the short-time spectra of a voiced utterance. Given speakers A and B their short-time spectra are approximately related as $X_A(\omega) = X_B(\alpha \cdot \omega)$ in case of the same utterance.

Several techniques for handling this warping effect have been proposed. One group of techniques tries to adapt the acoustic models to the features of the individual speakers, for example, [2,3]. These methods are also known as (constrained) MLLR techniques. Other methods try to normalize the spectral effects of different VTLs at the feature extraction stage [4,5] in order to reduce the mismatch between training and testing conditions. Though both groups of methods

---

* This work has been supported by the German Research Foundation under Grant No. ME1170/2-1.

J. Solé-Casals and V. Zaiats (Eds.): NOLISP 2009, LNAI 5933, pp. 111–119, 2010.

are working in different stages of an ASR system, it was shown in [3] that they are related by a linear transformation. In contrast to the mentioned techniques, a third group of methods avoids the additional adaption step within the ASR system by generating features that are independent of the warping factor [6,7,8,9].

A known approach for the time-frequency (TF) analysis of speech signals is to equally space the center frequencies of the filters on the quasi-logarithmic ERB scale. This scale approximately represents the frequency resolution of the human auditory system. Within this domain, linear frequency warping can approximately be described by a translation. On basis of the TF-analysis, this effect can be utilized for the computation of translation-invariant features [7,8,9].

Nonlinear transformations that lead to translation-invariant features have been investigated and successfully applied in the field of pattern recognition for decades. Following the concepts of [10], the general idea of invariant features is to find a mapping T that is able to extract features in such a way that they are the same for possibly different observations of the same equivalence class with respect to a group action. Thus, a transformation T maps all observations of an equivalence class into one point of the feature space. Given a transformation T, the set of all observations that are mapped into one point is denoted as the *set of invariants* of an observation. The set of all possible observations within one equivalence class is called *orbit*. A transformation T is said to be *complete*, if both the set of invariants of an observation and the orbit of the same observation are equal. Complete transformations have no ambiguities regarding the class discrimination. In practical applications, however, usually one has to deal with non-complete transformations.

The idea of the method proposed in this work is to extract features that are robust against VTL changes by using nonlinear transformations that are invariant to translations. Well-known transformations of this type are, for example, the cyclic autocorrelation of a sequence and the modulus of the discrete Fourier transformation (DFT). A general class of translation-invariant transformations was introduced in [11] and further investigated in [12,13] in the field of pattern recognition. The transformations of this class, which will be called $\mathbb{C}T$ in the following, can be computed efficiently.

Different transformations will be investigated in this paper with the aim of obtaining a feature set that has a high degree of completeness under the group action induced by VTL changes. Results will be presented for large-vocabulary phoneme recognition tasks with a mismatch in the mean VTL between the training and testing sets. Experiments show that the individual transformations of the class $\mathbb{C}T$ as well as previously investigated individual transforms [7,8] achieve a recognition performance that is comparable to the one of mel-frequency cepstral coefficients (MFCC). However, combinations of different invariant transformations significantly outperform the MFCCs with respect to the problem of VTL changes.

The next section introduces the class of transformations $\mathbb{C}T$ and explains our method for using these transformations for the extraction of features for speech recognition tasks. Section 3 describes the experimental setup. Results are presented in Section 4, followed by some conclusions in the last section.

2   Method

2.1   Translation-Invariant Transformations of Class $\mathbb{C}\bullet$

A general class of translation-invariant transformations was originally introduced in [11] and later given the name $\mathbb{C}T$ [12]. Their computation is based on a generalization of the fast Walsh-Hadamard transform (WHT). Given a vector $\bullet := (x_0, x_1, \ldots, x_{N-1})$ with $N = 2^M$ as input and following the notation of [10], members of the class $\mathbb{C}T$ are defined by the following recursive transformation T with commutative operators $f_1(.,.)$, $f_2(.,.)$:

$$T(\bullet) := (T(f_1(\bullet_1, \bullet_2)), \, T(f_2(\bullet_1, \bullet_2))) . \tag{1}$$

Herein, $\bullet_1$ and $\bullet_2$ denote the first and second halves of the vector $\bullet$, respectively. The recursion ends with $T(x_i) = x_i$. Fig. 1 shows a corresponding signal-flow diagram for $N = 4$.

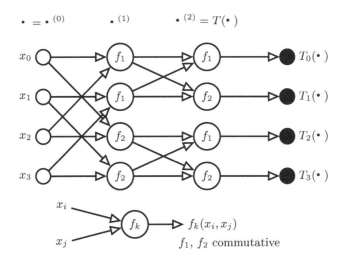

Fig. 1. Signal-flow diagram for transformations of the class $\mathbb{C}T$ for $N = 4$

The pairs of commutative operators that are examined in this work have found applications in pattern recognition tasks [11,12,14]. One representative of the class $\mathbb{C}T$ is the *rapid transformation* (RT) which has found a notably wide application [10,15,16]. In comparison to the RT, it was shown in [11] that taking the $\min(.,.)$ and $\max(.,.)$ functions as f_1 and f_2, respectively, can lead to better separability properties. The transformation with this pair of functions is denoted as "MT". It was shown in [12] that the power spectrum of the modified

Table 1. Common pairs of commutative operators

	RT	**MT**	**QT**
f_1	$a + b$	$\min(a, b)$	$a + b$
f_2	$\|a - b\|$	$\max(a, b)$	$(a - b)^2$

WHT can be computed with a transformation of $\mathbb{C}T$ by choosing $f_1 := a + b$ and $f_2 := (a - b)^2$. This transformation is denoted as "QT". The mentioned transformations together with their according pairs of commutative functions are summarized in Table 1.

In [13], a preprocessing operator for the RT was presented that destroys the unwanted property of invariance under reflection of the input data, and, thus, increases the separation capability of the RT. This operator, denoted as b, works element-wise and is defined as

$$x_i' = b(x_i, x_{i+1}, x_{i+2}) := x_i + |x_{i+1} - x_{i+2}| \; . \tag{2}$$

This particular preprocessing followed by the RT is called *modified rapid transformation* (MRT).

2.2 Translation-Invariant Feature Candidates for ASR

The translation-invariant features for an input signal $\bullet$ are computed on the basis of the magnitude of a TF-analysis. The TF representation will be denoted by $y_\bullet(t, k)$ in the following. Here, t is the frame index, $1 \le t \le T$, and k is the subband index with $0 \le k \le K - 1$. The transformations RT, MRT, MT, and QT are applied frame-wise to the primary features. In addition to the transformations described above, individual translation-invariant features from previous work [7,8] are also considered in this study. These are based on the logarithmized correlation sequences of spectral values,

$$\log r_{yy}(t, d, m) \quad \text{with} \quad r_{yy}(t, d, m) = \sum_k y_\bullet(t, k) y_\bullet(t - d, k + m) \tag{3}$$

and on the correlation sequences of logarithmized spectral values,

$$c_{yy}(t, d, m) = \sum_k \log(y_\bullet(t, k)) \cdot \log(y_\bullet(t - d, k + m)) \; . \tag{4}$$

Besides using the TF analysis $y_\bullet(t, k)$ directly as input for the transformations, we also consider multi-scale representations of it. The method of multi-scale analysis has been successfully applied to various fields of speech processing [17,18,19,20]. Multiple scales of spectral resolution for each frame were computed. The length of a frame on scale n is half the length of scale $n - 1$. Each scale was used as input to the described transformations, and the results of the transformations on each scale were concatenated. Following this procedure, the resulting number of features for an input of size $N = 2^M$ is $2^{M+1} - 1$. Features of this type are denoted with the subscript "Scales". For the experiments, also a subset of 50 features of the "Scales"-versions of the features was determined by applying a feature-selection method according to [21]. The feature-selection method uses a mutual information criterion, and the resulting feature sets are denoted with the subscript "Scales-50".

3   Experimental Setup

On the basis of the feature types described in Section 2, different feature sets have been defined and evaluated in a number of phoneme recognition experiments. The experiments have been conducted using the TIMIT corpus with a sampling rate of 16 kHz. To avoid an unfair bias for certain phonemes, we chose not to use the "SA" sentences in training and testing similar to [22]. Training and testing sets were both split into female and male utterances. This was used to simulate matching and mismatching training and testing conditions with respect to the mean VTL. Three different training and testing scenarios were defined: Training and testing on both male and female data (FM-FM), training on male and testing on female data (M-F), and training on female and testing on male data (F-M). According to [22], 48 phonetic models were trained, and the recognition results were folded to yield 39 final phoneme classes that had to be distinguished.

The recognizer was based on the Hidden-Markov Model Toolkit (HTK) [23]. Monophone models with three states per phoneme, 8 Gaussian mixtures per state and diagonal covariance matrices were used together with bigram language statistics.

MFCCs were used to obtain baseline recognition accuracies. The MFCCs were calculated by using the standard HTK setup which yields 12 coefficients and a single energy feature for each frame. For comparison with a vocal tract length normalization (VTLN) technique, the method of [4] was used.

We chose to use a complex-valued Gammatone filterbank [24] with 90 filters equally-spaced on the ERB scale as basis for computing the translation-invariant features. This setup was chosen to allow for a comparison with the previous works [7,8]. The magnitudes of the subband signals were low-pass filtered in order to decrease the time resolution to 20 ms. These filtered magnitudes were then subsampled to obtain a final frame rate of one frame every 10 ms. Because the transforms of the class $\mathbb{C}T$ require the length of the input data to be a power of two, the output of the filterbank was frame-wise interpolated to 128 data points. A power-law compression [25] with an exponent of 0.1 was applied in order to resemble the nonlinear compression found in the human auditory system.

The following feature types known from [7,8] were investigated in addition to class-$\mathbb{C}T$ features: The first 20 coefficients of the discrete cosine transform (DCT) of the correlation term (3) with $d = 0$ (denoted as "ACF") have been used, as well as the first 20 coefficients of the DCT of the term (4) with $d = 4$ (denoted as "CCF"). The features belonging to the class $\mathbb{C}T$ as described in the previous section were considered together with their "Scales" and "Scales-50" versions. In order to limit the size of the resulting feature vectors, the "Scales-50" versions were used for feature-set combinations of size four and five.

All feature sets were amended by the logarithmized energy of the original frames together with delta and delta-delta coefficients [23]. The resulting features were reduced to 47 features with linear discriminant analysis (LDA). The reduction matrices of the LDAs were based on the 48 phonetic classes contained in both the male and female utterances.

4   Results and Discussion

At first, all of the previously described feature types were tested individually in the three scenarios. The resulting accuracies of these experiments are shown in Table 2. It can be seen that the MFCCs have the highest accuracy for the FM-FM scenario compared to the other considered feature types. The features resulting from the RT and MRT obtain similar accuracies as the MFCCs in the mismatching scenarios, but perform worse in the matching scenario. The inclusion of different scales in the feature sets leads to accuracies that are comparable to those of the MFCCs in the FM-FM scenario and already outperform the MFCCs in the mismatching scenarios M-F and F-M. Using only the 50 best features from the "Scales"-feature sets leads to accuracies that are similar to the feature sets that include all scales. However, in the mismatching scenarios the "Scales-50" versions perform worse than the "Scales" version. The correlation-based features perform similar to the $\mathbb{C}T$-based ones. In comparison, the cross correlation features CCF perform better than the ACF features. This indicates the importance of contextual information for ASR.

As a further baseline performance, the VTLN method [4] using MFCCs as features has been tested on the three scenarios. Since this method adapts to the vocal tract length of each individual speaker, it gave the best performance in all cases. The results were as follows: FM-FM: 68.61%, M-F: 64.02%, F-M: 63.39%.

To investigate in how far the performance of the translation-invariant features can be increased through the combination of different feature types, all possible combinations of the "Scales"-versions of the features and the ACF and CCF features have been considered. These include feature sets of two, three, four, and five types of features. For each of these feature sets of different size, the results for the best combinations are shown in Table 3.

As the results show, the combination of two well-selected feature sets leads to an accuracy that is comparable to the MFCCs in the matching case. However, in contrast to the MFCCs, feature-type combinations lead to an accuracy that was 5.6% to 7% higher in the M-F scenario and 8.1% to 9.5% higher in the F-M scenario. In particular, the results indicate that the information contained in the CCF features is quite complementary to that contained in the $\mathbb{C}T$-based features. Also the MRT and MT features seem to contain complementary information. The observation that the accuracies do not increase by considering combinations of four or five feature sets could either be explained by the fact that the "Scales-50" features in comparison to the "Scales" features have a much

Table 2. Accuracies of individual feature types

Feature type	FM-FM	M-F	F-M
MFCC	66.57	55.00	52.42
RT	58.39	55.30	51.99
MRT	57.90	53.88	50.75
QT	53.00	48.03	46.12
MT	59.96	56.53	54.45
RT_{Scales}	64.29	57.36	56.67
MRT_{Scales}	64.27	58.90	58.42
QT_{Scales}	62.64	56.75	55.34
MT_{Scales}	64.05	58.79	58.02
$RT_{Scales-50}$	64.47	55.49	54.28
$MRT_{Scales-50}$	64.08	55.66	54.03
$QT_{Scales-50}$	62.25	53.07	52.15
$MT_{Scales-50}$	64.19	53.77	52.38
ACF	58.85	46.97	48.76
CCF	62.46	54.54	53.41

Table 3. Highest accuracies for feature sets with different sizes and energy amendment

Feature type combination + energy	FM-FM	M-F	F-M
MT_{Scales} + CCF	65.74	61.13	60.52
MRT + CCF	65.36	60.60	60.51
MRT_{Scales} + CCF + ACF	65.90	61.75	61.94
MRT_{Scales} + MT_{Scales} + CCF	65.71	62.01	61.94
$MRT_{Scales-50}$ + CCF + ACF + $RT_{Scales-50}$	66.01	61.27	60.59
$MRT_{Scales-50}$ + CCF + ACF + $RT_{Scales-50}$ + $QT_{Scales-50}$	65.94	61.77	61.17

lower accuracy for the gender separated scenarios or by the assumption that the RT, MRT and QT do contain similar information.

In a third experiment, we amended the previously considered, fully translation-invariant features with MFCCs, as this had been necessary to boost the performance with the method in [7,8]. The results of the experiment are shown in Table 4. It is notable that the MFCCs do not increase the accuracies significantly in the matching scenarios and increase only slightly in the mismatching scenarios. This means that the MFCCs do not carry much more additional discriminative information compared to the feature set combinations that solely consist of translation-invariant features.

Using the best feature set presented in the previous work [8] leads to the following accuracies: FM-FM: 65.70%, M-F: 60.75% and F-M: 59.90%. As expected, these results indicate a better performance in the gender separated scenarios than the MFCCs. However, the new translation-invariant feature sets presented in this paper perform even better and do not rely on the MFCCs.

Table 4. Highest accuracy for feature type combinations with different sizes and MFCC amendment

Feature type combination + MFCC	FM-FM	M-F	F-M
$\mathrm{MT_{Scales}}$ + CCF	65.58	61.62	61.46
$\mathrm{MRT_{Scales}}$ + CCF + ACF	66.45	62.08	62.40
$\mathrm{MRT_{Scales\text{-}50}}$ + CCF + ACF + $\mathrm{RT_{Scales\text{-}50}}$	66.34	61.90	61.23
$\mathrm{MRT_{Scales\text{-}50}}$ + CCF + ACF + $\mathrm{RT_{Scales\text{-}50}}$ + $\mathrm{QT_{Scales\text{-}50}}$	66.46	61.85	61.96

5   Conclusions

Vocal tract length changes approximately lead to translations in the subband-index space of time-frequency representations if an auditory-motivated filterbank is used. Well-known translation-invariant transformations that were originally proposed in the field of pattern recognition have been applied in this paper in order to obtain features that are robust to the effects of VTL changes. We showed that combining certain types of translation-invariant features leads to accuracies that are similar to those of MFCCs in case of matching training and testing conditions with respect to the mean VTL. For mismatched training and testing conditions, the proposed features significantly outperform the MFCCs. This may lead to significantly more robustness in scenarios in which VTLs differ significantly, as, for example, in children speech. Therefore, children speech and further feature optimization will be subject of future investigations on nonlinear feature-extraction methods. Also the combination with other invariant feature types will be investigated. Compared to the VTLN method, our features do not require any speaker adaptation and are therefore much faster to compute and to use than VTLN.

References

1. Benzeghiba, M., Mori, R.D., Deroo, O., Dupont, S., Erbes, T., Jouvet, D., Fissore, L., Laface, P., Mertins, A., Ris, C., Rose, R., Tyagi, V., Wellekens, C.: Automatic speech recognition and speech variability: a review. Speech Communication 49(10-11), 763–786 (2007)
2. Gales, M.J.F.: Maximum likelihood linear transformations for HMM-based speech recognition. Computer Speech and Language 12(2), 75–98 (1998)
3. Pitz, M., Ney, H.: Vocal tract normalization equals linear transformation in cepstral space. IEEE Trans. Speech and Audio Processing 13(5 Part 2), 930–944 (2005) (ausgedruckt)
4. Welling, L., Ney, H., Kanthak, S.: Speaker adaptive modeling by vocal tract normalization. IEEE Trans. Speech and Audio Processing 10(6), 415–426 (2002)
5. Lee, L., Rose, R.C.: A frequency warping approach to speaker normalization. IEEE Trans. Speech and Audio Processing 6(1), 49–60 (1998)
6. Umesh, S., Cohen, L., Marinovic, N., Nelson, D.J.: Scale transform in speech analysis. IEEE Trans. Speech and Audio Processing 7, 40–45 (1999)

7. Mertins, A., Rademacher, J.: Frequency-warping invariant features for automatic speech recognition. In: Proc. IEEE Int. Conf. Acoust., Speech, and Signal Processing, Toulouse, France, May 2006, vol. V, pp. 1025–1028 (2006)
8. Rademacher, J., Wächter, M., Mertins, A.: Improved warping-invariant features for automatic speech recognition. In: Proc. Int. Conf. Spoken Language Processing (Interspeech 2006 - ICSLP), Pittsburgh, PA, USA, September 2006, pp. 1499–1502 (2006)
9. Monaghan, J.J., Feldbauer, C., Walters, T.C., Patterson, R.D.: Low-dimensional, auditory feature vectors that improve vocal-tract-length normalization in automatic speech recognition. The Journal of the Acoustical Society of America 123(5), 3066–3066 (2008)
10. Burkhardt, H., Siggelkow, S.: Invariant features in pattern recognition – fundamentals and applications. In: Nonlinear Model-Based Image/Video Processing and Analysis, pp. 269–307. John Wiley & Sons, Chichester (2001)
11. Wagh, M., Kanetkar, S.: A class of translation invariant transforms. IEEE Trans. Acoustics, Speech, and Signal Processing 25(2), 203–205 (1977)
12. Burkhardt, H., Müller, X.: On invariant sets of a certain class of fast translation-invariant transforms. IEEE Trans. Acoustic, Speech, and Signal Processing 28(5), 517–523 (1980)
13. Fang, M., Häusler, G.: Modified rapid transform. Applied Optics 28(6), 1257–1262 (1989)
14. Reitboeck, H., Brody, T.P.: A transformation with invariance under cyclic permutation for applications in pattern recognition. Inf. & Control. 15, 130–154 (1969)
15. Wang, P.P., Shiau, R.C.: Machine recognition of printed chinese characters via transformation algorithms. Pattern Recognition 5(4), 303–321 (1973)
16. Gamec, J., Turan, J.: Use of Invertible Rapid Transform in Motion Analysis. Radioengineering 5(4), 21–27 (1996)
17. Pinkowski, B.: Multiscale fourier descriptors for classifying semivowels in spectrograms. Pattern Recognition 26(10), 1593–1602 (1993)
18. Stemmer, G., Hacker, C., Noth, E., Niemann, H.: Multiple time resolutions for derivatives of Mel-frequency cepstral coefficients. In: IEEE Workshop on Automatic Speech Recognition and Understanding, December 2001, pp. 37–40 (2001)
19. Mesgarani, N., Shamma, S., Slaney, M.: Speech discrimination based on multiscale spectro-temporal modulations. In: Proc. IEEE Int. Conf. Acoustics, Speech, and Signal Processing, May 2004, vol. 1, pp. I-601–I-604 (2004)
20. Zhang, Y., Zhou, J.: Audio segmentation based on multi-scale audio classification. In: IEEE Int. Con. Acoustics, Speech, and Signal Processing, May 2004, vol. 4, pp. iv-349–iv-352 (2004)
21. Peng, H., Long, F., Ding, C.: Feature selection based on mutual information: Criteria of max-dependency, max-relevance, and min-redundancy. IEEE Trans. Pattern Analysis and Machine Intelligence 27(8), 1226–1238 (2005)
22. Lee, K.F., Hon, H.W.: Speaker-independent phone recognition using hidden Markov models. IEEE Trans. Acoustics, Speech and Signal Processing 37(11), 1641–1648 (1989)
23. Young, S., Evermann, G., Gales, M., Hain, T., Kershaw, D., Liu, X.A., Moore, G., Odell, J., Ollason, D., Povey, D., Valtchev, V., Woodland, P.: The HTK Book (for HTK version 3.4). Cambridge University Engineering Department, Cambridge (2006)
24. Patterson, R.D.: Auditory images: How complex sounds are represented in the auditory system. Journal-Acoustical Society of Japan (E) 21(4), 183–190 (2000)
25. Bacon, S., Fay, R., Popper, A.: Compression: from cochlea to cochlear implants. Springer, Heidelberg (2004)

Time-Frequency Features Extraction for Infant Directed Speech Discrimination

Ammar Mahdhaoui, Mohamed Chetouani, and Loic Kessous

UPMC Univ Paris 06, F-75005, Paris, France CNRS, UMR 7222
ISIR, Institut des Systèmes Intelligents et de Robotique, F-75005, Paris, France
Ammar.Mahdhaoui@robot.jussieu.fr, Mohamed.Chetouani@upmc.fr

Abstract. In this paper we evaluate the relevance of a perceptual spectral model for automatic motherese detection. We investigated various classification techniques (Gaussian Mixture Models, Support Vector Machines, Neural network, k-nearest neighbors) often used in emotion recognition. Classification experiments were carried out with short manually pre-segmented speech and motherese segments extracted from family home movies (with a mean duration of approximately 3s). Accuracy of around 86% were obtained when tested on speaker-independent speech data and 87.5% in the last study with speaker-dependent. We found that GMM trained with spectral feature MFCC gives the best score since it outperforms all the single classifiers. We also found that a fusion between classifiers that use spectral features and classifiers that use prosodic information usually increases the performance for discrimination between motherese and normal-directed speech (around 86% accuracy).

Keywords: Automatic motherese discrimination, feature extraction, classification, parent-infant interaction.

1 Introduction

Infant-directed speech is a simplified language/dialect/register [2]. From an acoustic point of view, motherese has a clear signature (high pitch, exaggerated intonation contours) [1] [7]. The phonemes, and especially the vowels, are more clearly articulated. The exaggerated patterns facilitate the discrimination between the phonemes or sounds. In addition, motherese plays a major role since it is a highly communicative and social event in parents child communication that can elicit emotional reactions. Even if motherese is clearly defined in terms of acoustic properties, the modelling and the detection is expected to be difficult which is the case of the majority of emotional speech. Indeed, the characterization of spontaneous and affective speech in terms of features is still an open question and several parameters have been proposed in the literature [11]. In [8] [9], we found that prosodic features alone is not sufficient to characterize motherese, that's why we investigated spectral and cepstral features. Spectral or Cepstral (MFCC) coefficients are usually calculated on successive time windows of short duration. In a previous work [8] we presented a speaker-dependent

J. Solé-Casals and V. Zaiats (Eds.): NOLISP 2009, LNAI 5933, pp. 120–127, 2010.

infant-directed speech detection by investigation of two different classifiers and two different feature extraction techniques (segmental and supra-segmental). In this contribution we therefore aim to focus on speaker independent performance using multi classifiers and multi features techniques. The paper is structured as follows: Section 2 provides insights in feature extraction techniques and introduces the ensemble classification variants considered. Section 3 deals with experimental results and database description, and finally conclusions are drawn.

2   Method

2.1   Feature Extraction

As described in last study [8], [9], we processed two different features, segmental and supra-segmental features. Segmental features are characterized by the Mel Frequency Cepstrum Coefficients (MFCC), MFCC implies the use of a 'Mel scale' (closer to perception than linear frequency), it has been successfully used in speech and speaker recognition to represent the speech signal. We chose MFCC features to model the spectral properties of motherese and speech. Each 20 ms, 16 MFCC coefficients were computed with overlapping between adjacent frames equal to 1/2. The supra-segmental features are characterized by 3 statistics (mean, variance and range) of both F0 and short-time energy resulting on a 6 dimensional vector. One should note that the duration of the acoustic events is not directly characterized as a feature but it is taken into account during the classification process by a weighting factor [8]. The feature vectors are normalized (zero mean, unit standard deviation).

To improve our system in the case of speaker-independent, and according to the definition of motherese in the last section, we investigated the bark filters spectral representation, since motherese and normal speech perceptually sound different. The most useful parameters in speech processing are found in the frequency domain, because the vocal tract produces signals that are more consistently and easily analyzed spectrally than in the time domain [13], in addition, the features based on Bark scale are considered to provide more information by characterizing the human auditory system [13]. We extracted the bark time/frequency representation using an analysis window duration of 15 ms, a time step of 5 ms with filters equally spaced by 1 bark (first filter centered on first bark critical band) [6]. This computation on full spectrum, results on 29 filter bands. This representation can be described as a discrete perceptive representation of the energy spectrum which can be qualified as a perceptive spectrogram. We then extracted statistical features from this representation. Features are extracted either along the time axis or along the frequency axis. We also consider the average of energy of the bands (a perceptive Long Term Average Spectrum) and extract statistical features from it. 32 statistical features are used and applied a) along time axis, b) along frequency axes, c) on the average perceptive spectrum to obtain a first set of 32 features.

- Approach TL (for 'Time Line') Fig. 1. a.: (step 1) extracting the 32 features on the spectral vector of each time frame, then (step 2) averaging the values for each of 32 features along time, to finally obtain a second set of 32 features.
- Approach SL ('for Spectral Line') Fig. 1. b.: (step 1) extracting the 32 features along the time axis for each spectral band, (step 2) averaging the 32 features along the frequency axis to obtain a third set of 32 features.
- Approach MV (for 'Mean Values'): (step 1) averaging the energy values of the bark spectral bands along the time axis to obtain a long term average spectrum using 29 bark bands, (step 2) extracting the 32 statistical features from this average spectrum.

The method used for bark-based feature extraction is described in Fig. 1.

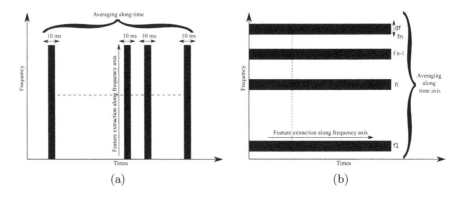

(a)   (b)

Fig. 1. Method for Extraction of bark-based features

The 32 statistical features we use are: maximum, minimum and mean value, standard deviation, variance, skewness, kurtosis, interquartile range, mean absolute deviation (MAD), MAD based on medians, i.e. MEDIAN(ABS(X-MEDIAN(X))), first and second coefficients of linear regression, first, second and third coefficients of quadratic regression, 9 quantiles corresponding to the following cumulative probability values: 0.025, 0.125, 0.25, 0.375, 0.50, 0.625, 0.75, 0.875, 0.975, quantile for cumulative probability values 0.1 and 0.9, and interquantile range between this two values, absolute and sign of time interval between maximum and minimum appearances. We use these statistics in order to models the dynamic variations of the bark spectral perceptive representation.

These 32 statistical features are also extracted from the pitch contour and the loudness contour. 3 other features are also extracted from these contours, by using a histogram and considering the maximum, the bin index of the maximum and the center value of the corresponding bin. These 3 features are relevant for pitch and energy contour.

2.2   Classification

In this subsection, we describe the different techniques used to model and discriminate motherese and speech employing the features described above. In addition to two classifiers, k-NN and GMM, described in [8], [9], we investigated Support Vector Machine (SVM) algorithm and neural networks (Multi Layer Perceptron).

Gaussian mixture model
Gaussian Mixture Model [10] is a statistical model which concerns modelling a statistical distribution of Gaussian Probability Density Functions (PDFs): a Gaussian Mixture Model (GMM) is a weighted average of several Gaussian PDFs. We trained 'motherese' GMMs and 'normal speech' GMMs with different sets of features. The GMMs were trained using the expectation maximization (EM) algorithm and with varying numbers of Gaussian mixtures (varying from 2 to 32 Gaussian mixtures for different feature sets) depending on the number of extracted features. In testing, a maximum likelihood criterion was used.

k-nearest neighbors
The k-NN classifier [4] is a distance based method. For the fusion process, we adopted a common statistical framework for the k-NN and the other classifiers by the estimation of a posteriori probabilities as described in [8] and [9].

Support vector machines
The Support Vector Machine (SVM) algorithm [12], is acknowledged to be a powerful method for 2-class discrimination and have become popular among many different types of classification problems, e.g., face identification, bioinformatics and speaker recognition. The basic principle of this discriminative method is to find the best separating hyperplane between groups of datapoints that maximizes the margins. We used LIBSVM [3], developed by Chih-Chung Chang and Chih-Jen Lin in the Departement of computer science and information engineerind of National Taiwan University, to model the SVMs using different sets of features, and tried several kernels (linear, Gaussian, polynomial and sigmoidal) that were available in this toolkit.

Neural network
Researchers from many scientific disciplined are designing Neural networks to solve a variety of problems in pattern recognition, prediction, optimization, associative memory, and control. The Neural Network structure which was used in this paper was the Multilayer Perceptron(MLP). The Weka data mining software is used as a toolbox for classification [5], to train and test our MLP classifiers.

3   Experimental Results

The performances of the GMM, SVM, NN and k-NN and fused classifiers are trained with different feature sets (MFCC, Pitch and Energy with statistics, bark features). To compare the performance of the different classifiers, we evaluated from a 10 folds cross validation the accuracy rate. To test our feature extraction methods with the different classifiers we use a real-life database that is described in the following section.

3.1   Database Description

In this study, we decided to look for a speech database that contains not acted and natural emotional speech. Furthermore, for practical reasons, the database should also include some paralinguistic or emotional annotation. Therefore, for training and testing, we decided to use a collection of natural and spontaneous interactions usually used for children development research (family home movies). This database meets our requirements: the corpus contains text-independent, speaker-independent, realistic, natural speech data and it contains human-made annotations of verbal interactions of the mother that's been carefully annotated by a psycholinguist on two categories: motherese and normal directed speech. From this manual annotation, we extracted 159 utterances for motherese and 152 for normal directed speech. The utterances are typically between 0.5s and 4s in length.

3.2   Results of Separate Classifiers

We started off training and testing GMM classifiers. All features are modeled using GMMs with diagonal covariance matrices measured over all frames of an utterance and we optimized the number of gaussian mixtures for each feature set. Table 1 shows that a GMM classifier trained with spectral MFCC features outperforms the GMM classifiers trained with the others features. A receiver operating characteristic (ROC) curve [4] of the best performing GMM classifier is shown in Fig. 2. The SVMs were further trained with the expanded features using a linear kernel, which is usually done in e.g., speaker recognition, the results of all the features used in SVMs are shown in Table 1, where we can observe that SVM using Bark features with first approach TL outperforms the other SVMs. The second-best performing feature set for SVM is the Bark features with all statistics. For K-nn as described in [8] we adjust the number of neighbors (k). Table 1 shows that k-NN with all statistics Bark features outperform the other features. For neural network we used Multi Layer Perceptron, In Table 1 we observe that MLP with all statistics of Bark features give the highest accuracy.

To summarize, comparing the results of the feature extraction techniques and taking into account the different classifiers, the best performing feature set for motherese detection appears to be spectral MFCC trained with GMM classifier (84%), the second best result is obtained with spectral Bark feature(74%). The performance of motherese detection depend not only on feature extraction method but also on classification technique.

3.3   Fusion

Since each of our separate classifiers were trained with a different set of features, it would be interesting to investigate whether using a combination of these classifiers would improve performance. We will focus on the fusion between various classifiers based on spectral MFCC features and the classifiers based on prosodic features. Fusions were carried out on score-level using fusion techniques described

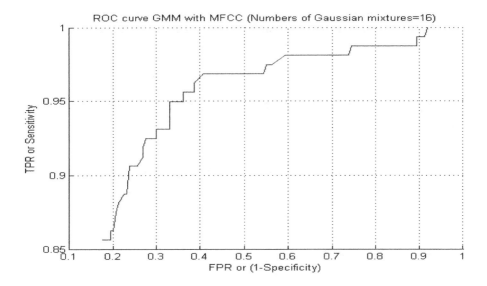

Fig. 2. ROC curves for GMM trained with spectral feature MFCC

Table 1. Accuracy of separate classifiers

	MFCC	Pitch & Energy (6 statistics)	Pitch & Energy (35 statistics)	Approach TL	Approach SL	Approach MV	Bark (96 statistics)
GMM	84%	67.5%	68.5%	69.5%	58.54%	67.71%	63.65%
K-nn	68.65%	70%	67%	71.38%	67.2%	69.78%	74 %
SVM	63.65%	63%	69%	74%	63.66%	69.77%	72.66%
MLP	72.8%	62%	67%	73%	67.85%	70.42%	74.60%

in [8]. We tested some fusion experiments and we show that whether the performances of the classifiers fused with MFCC and prosodic features are better than the performance of the single classifier trained with only spectral features MFCC since we obtained 86% accuracy as a result of fusion GMM trained with MFCC and GMM trained with prosodic, second best result of fusion is the combination between GMM trained with MFCC and SVM trained with bark features since we obtained 85.5% accuracy.

4   Conclusions

Our goal was to develop a motherese detector by computing multi-features and multi-classifiers in order to automatically discriminate pre-segmented motherese segments from manually pre-segmented speech segments to enable the study of parent infant interaction and investigation of the effect of this kind of speech on different children (typical vs autistic). The classification accuracies obtained in this study are very competitive with previously published results [8], [9] on the same database but here on speaker-independent. The results of this multi-classifiers and multi-features study underline the importance of performing more sophisticated performance measurements when evaluating supervised machine learning approaches with different features to the discrimination between infant-directed speech and normal-directed speech. By using more conventional features often used in speech/speaker recognition (spectral MFCC features) and other features prosodic and statistics on the pitch and energy and bark features in classification techniques, we were able to automatically discriminate motherese segments from normal speech segments. We obtained results from which we can draw interesting conclusions. Firstly, our results show that spectral features MFCC alone contain much useful information for discrimination between motherese and speech since they outperform all other features investigated in this study. Thus, we can conclude that spectral features MFCC alone can be used to discriminate between motherese and speech. However, according to our detection results, spectral bark features are also very promising features since without any features selection techniques we obtain the second-best accuracy.

For our motherese detection experiments, we used only segments that were already manually segmented (based on a human transcription) and segments that contained either motherese or speech. In other words, detection of onset and offset of motherese was not investigated in this study but can be addressed in a follow-up study. Detection of onset and offset of motherese (motherese segmentation) can be seen as a separate problem. Adding visual information could further help to improve the discrimination quality.

References

1. Andruski, J.E., Kuhl, P.K.: The acoustic structure of vowels in infant- and adult-directed speech. Paper presented at the Biannual Meeting of the Society for Research in Child Development, Washington, DC (April 1997)
2. Fernald, A., Kuhl, P.: Acoustic determinants of infant preference for Motherese speech. Infant Behavior and Development 10, 279–293 (1987)
3. Chang, C.-C., Lin, C.-J.: LIBSVM: a library for support vector machines (2001), http://www.csie.ntu.edu.tw/~cjlin/libsvm/
4. Duda, R., Hart, P., Stork, D.: Pattern Classification, 2nd edn. (2000)
5. Witten, E.F.I.H.: Data Mining: Practical Machine Learning Tools and Techniques with Java Implementations. The Kaufmann Series in Data Management Systems, Gray, J. Series (ed.) (October 1999)
6. Zwicker, E., Fastl, H.: Psychoacoustics: Facts and Models. Springer, Berlin (1999)
7. Fernald, A., Simon, T.: Expanded intonation contours in mothers speech to newborns. Developmental Psychology 20, 104–113 (1984)
8. Mahdhaoui, A., Chetouani, M., Zong, C.: Motherese Detection Based On Segmental and Supra-Segmental Features. In: International Conference on Pattern Recognition-ICPR, Tampa, Florida, USA, December 8-11 (2008)
9. Mahdhaoui, A., et al.: Automatic Motherese Detection for Face-to-Face Interaction Analysis. In: Esposito, A., et al. (eds.) Multimodal Signals: Cognitive and Algorithmic. Springer, Heidelberg (2009)

10. Reynolds, D.: ÒSpeaker identification and verification using Gaussian mixture speaker models. Ó Speech Communication 17, 91–108 (1995)
11. Schuller, B., Batliner, A., Seppi, D., Steidl, S., Vogt, T., Wagner, J., Devillers, L., Vidrascu, L., Amir, N., Kessous, L., Aharonson, V.: The relevance of feature type for the automatic classification of emotional user states: low level descriptors and functionals. In: Proceedings of Interspeech, pp. 2253–2256 (2007)
12. Vapnik, V.N.: The Nature of Statistical Learning Theory. Springer, New York (1995)
13. Zwicker, E.: Subdivision of the audible frequency range into critical bands. The Journal of the Acoustical Society of America 33 (February 1961)

Wavelet Speech Feature Extraction Using Mean Best Basis Algorithm

Jakub Gałka and Mariusz Ziółko

Department of Electronics, AGH University of Science and Technology, Al. Mickiewicza 30,
30-059 Kraków, Poland
{jgalka,ziolko}@agh.edu.pl

Abstract. This paper presents Mean Best Basis algorithm, an extension of the well known Best Basis Wickerhouser's method, for an adaptive wavelet decomposition of variable-length signals. A novel approach is used to obtain a decomposition tree of the wavelet-packet cosine hybrid transform for speech signal feature extraction. Obtained features are tested using the Polish language hidden Markov model phone-classifier.

Keywords: wavelet transform, best basis, speech parameterization, speech recognition.

1 Introduction

Almost all speech recognition systems transform acoustic waveforms into vectors that represent important features of speech signal. This process is called feature extraction or parameterization, and has been studied for a long time. Its aim is to reduce redundancy of representation of a signal without losing its content.

Mel-frequency cepstral coefficients (MFCC) and perceptual linear prediction (PLP) are the most popular and the most often used among other methods. These methods are based on algorithms developed from windowed discrete Fourier transform (DFT). Its main disadvantage is caused by an equal window size applied to each of various analyzed frequencies. The same time-resolution is used to measure different frequencies (too high or too low). It is inadvisable and may lead to noticeable propagation of border effect for some frequencies, followed by time-resolution loss for others, also when psychoacoustic mel-scale had been applied.

A wavelet transform performs an analysis of various frequencies (related to wavelet scales) using various and adequate windows lengths, therefore abovementioned disadvantages can be reduced. Classic discrete decomposition schemes, which are dyadic (DWT), and packet wavelet (WP), do not fulfill all essential conditions required for direct use in parameterization. DWT do not provide sufficient number of frequency bands for effective speech analysis; however it is a good approximation of the perceptual frequency division [1], [2]. Wavelet packets do provide enough frequency bands, however they do not respect the non-linear frequency perception phenomena [3], [4], [5].

J. Solé-Casals and V. Zaiats (Eds.): NOLISP 2009, LNAI 5933, pp. 128–135, 2010.

Various decomposition schemes for an efficient speech parameterization had been presented [6]. Most of publications present an approximation of perceptual frequency division with an arbitrary or empirically chosen decomposition subtree [7], [8], [9], [10], [11], [12]. These papers do not provide description of the subtree selection method. In some works wavelet filters have been warped or wavelet a-scale has been properly chosen to obtain mel-frequency scale in a wavelet transform [13].

Wickerhouser's best wavelet basis selection, entropy-based algorithm [14] has been used by Datta and Long [6] to obtain the best decomposition schemes of single phonemes. Other works mention use of this algorithm in a parameterization of plosive consonants [15].

Unfortunately, the well known Best Basis (BB) and Joint Best Basis (JBB) algorithms can not be used for sets of variable length data. In this paper, a new method of best wavelet basis selection is presented. It is applicable to sets of a non-uniform data, like various-length phoneme samples.

2 Mean Best Basis Decomposition

2.1 Wavelet Packet Cosine Transform (WPCT)

Multi-level wavelet packets produce 2^M wavelet coefficient vectors, where M stands for the number of decomposition levels. Wavelet coefficient vectors

$$\boldsymbol{d}_{m,j} = \left[d_{m,j} \right] \in \Re^K \tag{1}$$

represent uniformly distributed frequency banks. Decomposition process may be represented by a full binary tree

$$\boldsymbol{W}^{WPT} = \left\{ W_{m,j}^{WPT} \right\}: \quad W_{m,j}^{WPT} \leftrightarrow \boldsymbol{d}_{m,j} \tag{2}$$

with a sample of speech signal $\boldsymbol{d}_{0,0}$ (single frame of speech) related to its root, and wavelet coefficients $\boldsymbol{d}_{m,j}$ related to its nodes and leafs (when $m=M$) [16], [17].

For a better spectral entropy extraction from the speech signal we applied the discrete cosine transform

$$\hat{d}_{m,j}(k) = \sum_{n=1}^{N_m} d_{m,j}(n) \cdot \cos\left(2\pi \frac{nk}{N_m} \right) \tag{3}$$

to each of the WP tree nodes to obtain the Wavelet Packet Cosine Transform (WPCT). It eliminates the problem of time-shift in the entropy measure of the signal and takes account of more important spectral content for further best basis selection. This is a very important step since the speech is a time-spectral phenomenon [3], [4].

2.2 Best Basis Algorithm

The best wavelet basis subtree $\boldsymbol{W}^{opt}$ may be defined as a set $\boldsymbol{W}^*$ of tree nodes

$$\boldsymbol{W}^{opt} = \operatorname*{argmin}_{\boldsymbol{W}^*} \sum_{\varkappa_{m,j} \leftrightarrow \boldsymbol{W}^*} \varkappa_{m,j} \tag{4}$$

which minimizes its total entropy and generates an orthogonal decomposition base [14], where the node split cost function

$$\varkappa\left(\widehat{\boldsymbol{d}}_{m,j}\right) = -\sum_{n=1}^{N_d}\left(\frac{\widehat{d}_{m,j}^{\;2}(n)}{\left\|\widehat{\boldsymbol{d}}_{m,j}\right\|^2}\log\left(\frac{\widehat{d}_{m,j}^{\;2}(n)}{\left\|\widehat{\boldsymbol{d}}_{m,j}\right\|^2}\right)\right) \tag{5}$$

is the Shannon entropy of the WPCT coefficients.

The best basis algorithm may be applied to a single signal when it is needed. However, finding the best decomposition scheme for a set of signals can not be done using this method. When a set of signals is given, joint best basis algorithm may be used [18], [19]. It utilizes a tree of signal variances

$$\boldsymbol{W}^{\sigma,opt} = \underset{\boldsymbol{W}^{\sigma}}{\operatorname{argmin}} \sum_{\varkappa_{m,j}^{\sigma} \leftrightarrow \boldsymbol{W}^{\sigma}} \varkappa_{m,j}^{\sigma} \tag{6}$$

to select an optimized subtree. Unfortunately, computation of variance requires each signal to be of equal length and normalized in terms of energy and amplitude, what is even more important, when energy dependent cost function is used [14]. This is a serious limitation, since in practice signals may be of various lengths. The next section presents the solution of this problem by calculation of mean entropy values instead of signals' variances.

2.3 Mean Best Basis Algorithm

The set of speech signals used in this work consists of phoneme samples extracted from Polish speech database *CORPORA*. Phonemes are of various lengths, depending on the phoneme class and case. Actually each pattern is unique. Under this conditions the use of variance-based JBB algorithm is impossible. The tree of variances cannot be fairly computed when signals are of various lengths and energies [18].

The above-mentioned problem may be solved when a new definition of the optimal tree for a set of different signals is introduced. The best decomposition tree in such case is a subtree

$$\overline{\boldsymbol{W}}^{opt} = \underset{\overline{\boldsymbol{W}}^*}{\operatorname{argmin}} \sum_{\overline{\varkappa}_{m,j} \leftrightarrow \overline{\boldsymbol{W}}^*} \overline{\varkappa}_{m,j} \tag{7}$$

of a full binary tree $\overline{\boldsymbol{W}}^{\varkappa}$ of nodes' entropy mean values $\{\overline{\varkappa}\}$ over all signals in the set, for which its entire value is minimal. Having a tree of mean entropy values, one can find an optimal Mean Best Basis (MBB) subtree using the BB algorithm over mean entropy tree. The algorithm consists of the following steps:

1) For each element of set $\{s\}_i$ of signals calculate full WPCT tree

$$\boldsymbol{W}^{WPCT} = \{W_{m,j}^{WPCT}\} : \quad W_{m,j}^{WPCT} \leftrightarrow \widehat{\boldsymbol{d}}_{m,j} \;. \tag{8}$$

2) Find entropy value

$$\varkappa_{m,j}^i = \varkappa\left(\widehat{\boldsymbol{d}}_{m,j}^i\right) \tag{9}$$

for each node of all previously calculated WPCT trees.

3) For each of the obtained trees $W_i^{\varkappa}$ normalize entropy values within the whole tree according to its root entropy value

$$\forall_{i} \forall_{m,j} \; \varkappa_{m,j}^{i} = \frac{\varkappa_{m,j}^{i}}{\varkappa_{0,1}^{i}} \; . \tag{10}$$

It makes the cost-function (entropy) independent of different signal energy values. After this step, every signal from the set will be equally important in the basis selection process.

4) Calculate

$$\overline{W}^{\varkappa} = \{\overline{W}_{m,j}^{\varkappa} \leftrightarrow \overline{\varkappa}_{m,j}\}: \; \overline{\varkappa}_{m,j} = \frac{1}{\left|\{s\}\right|} \sum_{\varkappa_{m,j}^{i} \leftrightarrow W_{i}^{\varkappa}} \varkappa_{m,j}^{i} \; , \tag{11}$$

which is the general tree of mean entropy values over all signals with all entropy values normalized.

5) Find the best subtree using the Wickerhouser's Best Basis algorithm with a mean-entropy tree $\overline{W}^{\varkappa}$.

Obtained wavelet decomposition scheme depends on the entropy and spectral properties of all signals used in the computations. Frequency bands containing more spectral variations among all signals in the set are represented in the optimized wavelet spectrum with a higher spectral resolution.

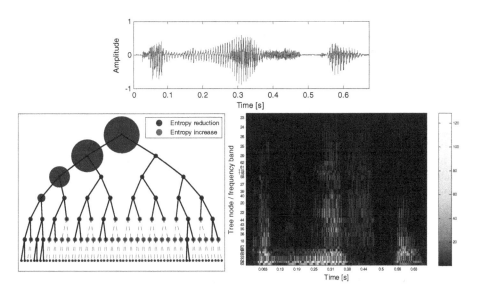

Fig. 1. Optimized MBB wavelet decomposition tree for polish speech database *CORPORA*, using Daubechies wavelet and Shannon entropy (*solid lines, left plot*). Utterance *"Agn'jeSka"* (SAMPA notation, *top*) and its MBB optimized spectrum (*right*).

In Fig. 1 a wavelet decomposition tree, obtained for all of the phones of Polish language with a Daubechie's 6th order wavelet and MBB algorithm is presented. The order of tree branches is not frequency-based because of the disordering effect of multilevel decimation / filtering present in the decomposition process [20]. In Fig. 1 it is also possible to notice a higher resolution of the spectrum in the frequency ranges related to the 1st and the 2nd formant. The spectrum has been generated using the tree presented in the left plot. Bands in the spectrum plot are frequency-ordered.

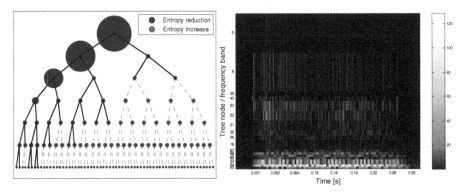

Fig. 2. Optimized MBB wavelet decomposition tree for Polish vowels, using Daubechies wavelet and Shannon entropy (*left*). MBB vowels-optimized wavelet spectrum of the phoneme /e/ (*right*).

2.4 Feature Extraction

When the optimized decomposition tree $\overline{W}^{opt}$ is known, it may be used for an efficient spectral analysis and feature extraction [8]. In presented experiment, energy

$$x(k) = \sum_{d_{m_k, j_k} \leftrightarrow W^{opt}} \left\| d_{m_k, j_k} \right\|^2 \tag{12}$$

of wavelet coefficient in each leaf was computed. Obtained values form a vector x of a length equal to the optimized tree's leafs quantity. Normalization and DCT decorelation of the vector is then applied to use it with a Hidden Markov Model (HMM) phone recognizer.

3   Phoneme Recognition

New decomposition schemes were tested using Polish speech database *CORPORA*. Phone recognition task had been performed using 3617 patterns. All phoneme patterns were used in the mean best basis selection. Obtained decomposition subtree had been used for speech feature extraction. In this case 27 tree leafs produced 27 features.

Its efficacy was measured with typical HMM tri-phone classifier with no higher-level language context knowledge [21]. Various noise conditions (Additive White Gaussian Noise, AWGN) had been applied to measure the robustness of the features.

Results of this task are presented in Fig. 3. For the given feature quantity (27), phone recognition and phone accuracy rates are reaching 80% and 72% respectively on clean speech. Introduction of 10dB SNR noise results in the recognition decrease by only 10% points which proves robustness of such composed wavelet parameterization scheme. Similar recognition task run on the vowels set with only 17 feature components resulted in 90% phone recognition accuracy for clean conditions with similar HMM setup.

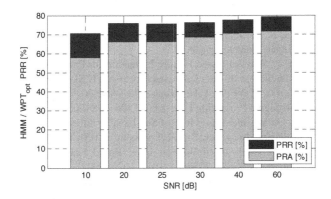

Fig. 3. Phoneme recognition results for the MBB-optimized parameterization scheme

4   Conclusions

A new method of choosing the best wavelet decomposition scheme for a set of signals has been presented. It is based on the well known Wickerhouser's BB algorithm, but extends it with the possibility of selecting the decomposition tree for differentiated multi-length data. The use of a WPCT, provides high robustness of the entropy value to a time-shift and focuses on the spectral properties of the signal. Decomposition schemes obtained for the real speech data and phone recognition results confirm the method's efficacy. Presented algorithm may be used with other types of signals, e. g. image data.

Future works will focus on finding the better, aim-oriented cost function (in place of entropy) used in a tree selection process.

Acknowledgments. This work was supported by MNiSW grant OR00001905. We would like to thank Stefan Grocholewski from Poznań University of Technology for providing a corpus of spoken Polish - *CORPORA*.

References

1. Datta, S., Farooq, O.: Phoneme Recognition Using Wavelet Based Features. An International Journal on Information Sciences 150 (2003)
2. Tan, B.T., Fu, M., Spray, A., Dermody, Ph.: The Use of Wavelet Transforms in Phoneme Recognition. In: 4th International Conference on Spoken Language Processing ICSLP (1996)
3. Gałka, J., Kepiński, M., Ziółko, M.: Speech Signals in Wavelet-Fourier Domain. In: 5th Open Seminar on Acoustics - Speech Analysis, Synthesis and Recognition In Technology, Linguistics And Medicine. Archives of Acoustics, vol. 28(3) (2003)
4. Gałka, J., Kepiński, M.: WFT context-sensitive speech signal representation. In: Kłopotek, M.A., Wierzchoń, S.T., Trojanowski, K. (eds.) IIPWM 2006. Advances in Soft Computing, pp. 97–105. Springer, Heidelberg (2006)
5. Ganchev, T., Siafarikas, M., Fakotakis, N.: Speaker Verification Based on Wavelet Packets. In: Sojka, P., Kopeček, I., Pala, K. (eds.) TSD 2004. LNCS (LNAI), vol. 3206, pp. 299–306. Springer, Heidelberg (2004)
6. Datta, S., Long, C.J.: Wavelet Based Feature Extraction for Phoneme Recognition. In: 4th International Conference on Spoken Language Processing ICSLP (1996)
7. Datta, S., Farooq, O.: Mel Filter-Like Admissible Wavelet Packet Structure for Speech Recognition. IEEE Signal Processing Letters 8(7), 196–198 (2001)
8. Datta, S., Farooq, O.: Wavelet Based Robust Sub-band Features for Phoneme Recognition. IEE Proceedings: Vision, Image and Signal Processing 151(3), 187–193 (2004)
9. Datta, S., Farooq, O.: Mel-Scaled Wavelet Filter Based Features for Noisy Unvoiced Phoneme Recognition. In: ICSLP 2002, pp. 1017–1020 (2002)
10. Gowdy, J.N., Tufekci, Z.: Mel-Scaled Discrete Wavelet Coefficients for Speech Recognition. In: 25th IEEE International Conference on Acoustics, Speech, and Signal Processing - ICASSP 2000, vol. 3, pp. 1351–1354. IEEE Press, New York (2000)
11. Sarikaya, R., Hansen, J.H.L.: High Resolution Speech Feature Parameterization for Monophone – Based Stressed Speech Recognition. IEEE Signal Processing Letters 7(7), 182–185 (2000)
12. Sarikaya, R., Gowdy, J.N.: Subband Based Classification of Speech Under Stress. In: 23rd IEEE International Conference on Acoustics, Speech, and Signal Processing – ICASSP 1998, vol. 1, pp. 569–572. IEEE Press, New York (1998)
13. Evangelista, G., Cavaliere, S.: Discrete Frequency Warped Wavelets: Theory and Applications. IEEE Transactions on Signal Processing 46(4), 874–875 (1998)
14. Wickerhauser, M.V., Coifman, R.R.: Entropy-Based Algorithms for Best Basis Selection. IEEE Transactions on Information Theory 38(2), part 2, 713–718 (1992)
15. Łukasik, E.: Classification of Voiceless Plosives Using Wavelet Packet Based Approaches. In: EUSIPCO 2000, pp. 1933–1936 (2000)
16. Daubechies, I.: Ten Lectures on Wavelets. SIAM, Philadelphia (1992)
17. Vetterli, M., Ramchandran, K., Herley, C.: Wavelets, Subband Coding, and Best Bases. Proceedings of the IEEE 84(4), 541–560 (1996)
18. Wickerhauser, M.V.: Designing a Custom Wavelet Packet Image Compression Scheme with Applications to Fingerprints and Seismic Data. In: Perspectives in Mathematical Physics: Conference in Honor of Alex Grossmann, pp. 153–157. CFML, CRC Press (1998)

19. Wickerhauser, M.V., Odgaard, P.F., Stoustrup, J.: Wavelet Packet Based Detection of Surface Faults on Compact Discs. In: 6th IFAC Symposium on Fault Detection, Supervision and Safety of Technical Processes. IFAC, vol. 6, part 1 (2006)
20. Vetterli, M., Ramchandran, K.: Best Wavelet Packet Bases Using Rate-Distortion Criteria. In: IEEE International Symposium on Circuits and Systems – ISCAS 1992, p. 971. IEEE Press, New York (1992)
21. Young, S., et al.: HTK Book. Cambridge University Engineering Department, Cambridge (2005)

Perceptually Motivated Generalized Spectral Subtraction for Speech Enhancement

Novlene Zoghlami[1], Zied Lachiri[1,2], and Noureddine Ellouze[1]

[1] Ecole Nationale d'Ingénieurs de Tunis / Département Génie Electrique, unité Signal,
Image et Reconnaissance de formes
ENIT, BP.37, Le Belvédère, 1002, Tunis, Tunisia
novlene_zoglami@yahoo.fr, N.Ellouze@enit.rnu.tn
[2] Institut National des Sciences Appliquées et de Technologie /
Département Instrumentation et Mesures
INSAT, BP 676 centre urbain cedex, Tunis, Tunisia
zied.lachiri@enit.rnu.tn

Abstract. This paper addresses the problem of single speech enhancement in
adverse environment. The common noise reduction techniques are limited by a
tradeoff between an efficient noise reduction, a minimum of speech distortion
and musical noise.in this work, we propose a new speech enhancement approach based on non-uniform multi-band analysis. The noisy signal is divided
into a number of sub-bands using a gammachirp filter bank with non-linear
ERB resolution, and the sub-bands signals are individually weighted according
the generalized spectral subtraction technique. For evaluating the performance
of the proposed speech enhancement, we use the perceptual evaluation measure
of speech quality PESQ and the subjective quality rating designed to evaluate
speech quality along three dimensions: signal distortion, noise distortion and
overall quality. Subjective evaluation tests demonstrate significant improvements results over classical subtractive type algorithms, when tested with
speech signal corrupted by various noises at different signal to noise ratios.

Keywords: speech enhancement, auditory filter bank, power spectral subtraction, noise estimation, and musical noise.

1 Introduction

With the increasing attractiveness of automatic speech processing systems, a need to
develop acoustic noise suppression rules for speech signal is imposed, since these
systems are often used in environment where high ambient noise levels are present, so
their performance degrades considerably. This problem has already received much
attention in the literature, and many algorithms are developed in order to removing
the background noise while retaining speech intelligibility. Actually, noise suppression algorithms are based on short-time spectral estimation. These methods are relied
with a tradeoff between a minimum level of speech distortion introduced and efficient
noise suppression. The spectral subtraction rule is the most popular technique able to
reduce the background noise using estimation   of the short-time spectral magnitude

J. Solé-Casals and V. Zaiats (Eds.): NOLISP 2009, LNAI 5933, pp. 136–143, 2010.

of the speech signal, obtained by subtracting the noise estimation from the noisy speech. However, this method needs to be improved, since it introduces in the enhanced speech a perceptually annoying residual noise, called musical noise, and composed of tones and random frequencies. To overcome this problem, the Ephraim and Malah subtraction rule [2] exploits the average spectral estimation of the speech signal based on a prior knowledge of the noise variance, in the goal to mask and reduce the residual noise. Others methods [3], [4] exploit the proprieties of the human auditory system, especially the auditory masking to improve the quality and intelligibility of the speech signal without introducing speech distortion. The objective of this paper is to adapt the generalized spectral subtraction technique to a multi-bands analysis using non-linear frequency ERB resolution filterbank, in according with the human auditory system behavior. In section 2, we present the subtractive type algorithms. Section 3 shows the proposed noise reduction approach based on auditory spectral analysis and the section 4 exposes the results and subjective evaluation tests.

2 Speech Enhancement Based on Perceptual Filterbank

The generalized spectral subtraction [1] is performed in noise suppression using short time spectral analysis based on fixed and uniform spaced frequency transformation. A critical problem is often appears when the processed speech is degraded by unnatural fluctuations generated via spectral gain function used in the subtractive process. These fluctuations caused by the variation of the noise spectrum estimate, are perceived as a varying artifacts that observed such a small random peaks distributed over the time frequency plane. These artifacts are described as residual noise known as musical noise which consists of tonal remnant noise component significantly disagreeable to the ear. The annoying fluctuations and the transition component in the structure of the musical noise depend on temporal and frequency speech analysis. Due to this fact, several perceptual based approach instead of completely eliminate the musical noise. In fact, many aspects of audiology and psychoacoustics demonstrate that the human auditory system may be sensitive to abrupt artifacts changes and transient component in the speech signal based on time-frequency analysis with a non linear frequency selectivity of the basilar membrane. Thus the human ear analysis can be conceptualized as an array of overlapping band-pass filters knows as auditory filters. These filters occur along the basilar membrane and increase the frequency selectivity of the human ear. According these assumptions, an improvement speech perception in noise environment may be possible, since the speech component can be identified and the selectivity can be amplified. This suggests an approach to improving the quality and intelligibility of speech in background noise using spectral analysis according psychoacoustics aspects. In order to enhance the noisy speech and avoid efficiently the musical noise, it is interesting to implement the generalized spectral subtraction method according a non uniform filterbank analysis.

2.1 Auditory Filters Modelling

The aim in auditory modeling is to find mathematical model which represents some physiological and perceptual aspects of the human auditory system. Auditory modeling is very useful, since the sound wave can be analyzed according the human ear comportment, with a good mode.

The simplest way to model the frequency resolution of the basilar membrane is to make analysis using filterbank. The adaptive and the most realistic model is the gammachirp filterbank [12], it is an extension of the popular gammatone filter [5] with an additional frequency modulation term to produce an asymmetric amplitude spectrum. The complex impulsion response is based on psychoacoustics measurements, providing a more accurate approximation to the auditory frequency response, and it is given in the temporal model as:

$$g(t) = At^{n-1} \exp(-2\pi bB(f_c)t) \cos(2\pi f_c t + c \ln t + \varphi) \tag{1}$$

where time $t>0$, A is the amplitude, n and b are parameters defining the envelope of the gamma distribution, f_c is the asymptotic frequency, c is a parameter for the frequency modulation ($c=3$), φ is the initial phase, $\ln t$ is a natural logarithm of time, and ERB (f_c) is the equivalent rectangular bandwidth of the auditory filter at f_c.

2.2 Choice of Frequency Scale

The frequency resolution of human hearing is a complex phenomenon which depends on many factors, such as frequency, signal bandwidth, and signal level. Despite of the fact that our ear is very accurate in single frequency analysis, broadband signals are analyzed using quite sparse frequency resolution. The Equivalent Rectangular Bandwidth (ERB) scale is an accurate way to explain the frequency resolution of human hearing with broadband signals. The expression used to convert a frequency f in Hz in its value in ERB is:

$$ERB(f) = 21.4 \log(\frac{4.37 f}{1000} + 1) \tag{2}$$

2.3 Perceptual Based Spectral Subtraction

The proposed speech enhancement method (figure1) is based on non-uniform decomposition of the input waveform $y(n)$. The processing is done by dividing the incoming noisy speech into separate bands $y_s(n)$ that could be individually manipulated using the spectral subtractive algorithm to achieve quality and intelligibility improvement of the overall signal.

The analysis filterbank consists of 4th order gammachirp filters that cover the frequency range of the signal. The filters bandwidth changes according the Equivalent rectangular bandwidth ERB scale. The output of the sth filter of the analysis filterbank can be expressed as:

$$y_s(n) = y(n) * g_s(n) \tag{3}$$

Where $g_s(n)$ is the impulse response of the sth, 4th-order gammachirp filter.

Each sub-band of the noisy speech $y_s(n)$ is manipulated after windowing (hamming window) using the spectral gain $G_s^{ss}(k,p)$ given by the generalized spectral subtraction rule in each frame p and each frequency k. The enhanced speech spectrum in each sub-band $|\hat{X}_s(k,p)|$ is given by:

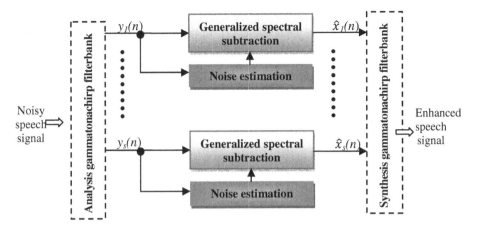

Fig. 1. Proposed speech enhancement method based on perceptual filterbank

$$\left|\hat{X}_s(k, p)\right| = G_s^{ss}(k, p) \cdot \left|Y_s(k, p)\right| \tag{4}$$

Where the gain function $G_s^{ss}(k, p)$ is expressed in each sub-band s as:

$$G_s^{ss}(k, p) = \begin{cases} \left(\left(1 - \alpha\left[\dfrac{\left|\hat{N}_s(k, p)\right|}{\left|Y_s(k, p)\right|}\right]^2\right)\right)^{1/2} & if\ \left|\hat{X}_s(k,p)\right|^2 > \beta\left|\hat{N}_s(k, p)\right|^2 \\[2em] \left(\beta\left|\hat{N}_s(k, p)\right|^2\right)^{1/2} & otherwise. \end{cases} \tag{5}$$

The parameters of the subtraction rule are set to ($\alpha=1$ and $\beta=0,0002$).

$\left|\hat{X}_s(k, p)\right|^2$; $\left|Y_s(k, p)\right|^2$ and $\left|\hat{N}_s(k, p)\right|^2$ are the kth power spectrrum, in each frame p and each sub-band s of the clean speech, noisy speech and estimate noise spectrum. The temporal enhanced speech signal $\hat{x}_s(n)$ in each temporal sub-band is estimated using the overlap-add technique and the inverse Fourier transform. In the synthesis filterbank, the final enhanced output speech signal $\hat{x}(n)$ is obtained using the summation of the sub-band signals after processing:

$$\hat{x}(n) = \sum_s \hat{x}_s(n) \tag{6}$$

The noise estimate has an important impact on the quality and intelligibility of the enhanced signal. If the noise estimate is too low, a residual noise will be audible, if the noise estimate is too high; speech will be distorted resulting in intelligibility loss. In the spectral subtraction algorithm, the noise spectrum estimate is update during the silent moment of the signal. Although this approach might give satisfactory result with stationary noise, but it's a problem with more realistic environments where the spectral characteristics of the noise change constantly. Hence there is a need to update the noise spectrum continuously over time. Several noise-estimation algorithms have been proposed for speech enhancement applications [7]. In [8], the method for estimating the noise spectrum (Martin) is based on tracking the minimum of the noisy speech over a finite window based on the statistics of the minimum estimates. In [9], a minima controlled recursive algorithm (MCRA) is proposed; it updates the noise estimate by tracking the noise-only regions of the noisy speech spectrum. In the improved minima controlled recursive algorithm (IMCRA) approach [10], a different method was used to track the noise-only regions of the spectrum based on the estimated speech-presence probability. Recently a new noise estimation algorithm (MCRA2) was introduced [11], the noise estimate was updated in each frame based on voice activity detection based on the ratio of noise speech spectrum to its local minimum.

3 Results and Evaluation

In our work, we evaluate the proposed auditory spectral attenuation for speech enhancement compared with the power spectral subtraction and the Ephraim and Malah subtractive rules under a noise environment. Speech signals are taken from a noisy speech corpus (NOIZEUS) sampled at 8 KHz and degraded by different noises, suburban train noise, multi-talker babble, car and station noise. The test signals include 30 speech utterances from 3 different speakers, female and male. To cover the frequency range of the signal, the analysis stage used in the auditory subtraction consists of 27- 4th order gammachirp filters according to ERB scale. The background noise spectrum is continuously estimated using the MCRA, IMCRA, MCRA2 and Martin noise estimation algorithms. A part from noise reduction, naturalness and intelligibility of enhanced speech are important attributes of the performance of any speech enhancement system. Since achieving a high degree of noise suppression is often accompanied by speech signal distortion, it is important to evaluate both quality and intelligibility. The performance evaluation in our work includes a subjective test of perceptual evaluation of speech quality PESQ [6].In fact, significant gains in noise reduction are accompanied by a decrease in speech intelligibility. Formal subjective listening test are the best indicates of achieved of overall quality. So the subjective listening test [7] used instructs the listener to successively attend and rate the enhanced speech signal on : the speech signal alone using five-point scale of signal distortion (SIG), the background noise alone using a five-point scale of background intrusiveness (BAK) and the overall effect using the scale of the mean opinion score (OVRL) [1=bad, 2=poor, 3=fair, 4=good, 5=excellent]. This process is designed to integrate the effects of both the signal and the background in making the rating of overall quality.

Table 1. Scale of signal distortion (SIG) and scale of background intrusiveness (BAK).

Scores	SIG	BAK
5	Not noticeable	very natural, no degraded
4	Somewhat noticeable	Fairly natural, little degradation
3	noticeable but not intrusive	Somewhat natural, somewhat degraded
2	Fairly conspicuous, somewhat intrusive	Fairly unnatural, fairly degraded
1	Very conspicuous, very intrusive	very unnatural, very degraded

Table 2. PESQ score for the proposed perceptual generalized spectral subtraction method (PGSS) compared with the spectral subtraction *(GSS)*.

	SNR	GSS	PGSS MCRA	PGSS IMCRA	PGSS MCRA2	PGSS MARTIN
Babble	0dB	1.78	1.83	1.84	1.82	1.81
	5dB	2,10	2.17	2.17	2.14	2.10
	10dB	2,40	2.51	2.51	2.47	2.42
	15dB	2,73	2.87	2.85	2.81	2.76
Car	0dB	1,83	1.93	1.90	1.91	1.83
	5dB	2,12	2.25	2.22	2.20	2.09
	10dB	2,46	2.59	2.56	2.54	2.41
	15dB	2,73	2.93	2.93	2.88	2.74
Station	0dB	1,73	1.86	1.87	1.84	1.81
	5dB	2,07	2.24	2.25	2.21	2.14
	10dB	2,40	2.58	2.59	2.52	2.43
	15dB	2,65	2.88	2.89	2.84	2.74
train	0dB	1,81	1.84	1.86	1.83	1.75
	5dB	2,17	2.14	2.12	2.11	2.01
	10dB	2,45	2.40	2.44	2.41	2.31
	15dB	2,72	2.85	2.84	2.80	2.67

Table 3. SIG- BAK -OVRL scores for the proposed perceptual generalized spectral subtraction method (PGSS) compared to the subtractive algorithms at 0 and 5 dB

SNR= 0dB		Babble SIG	Babble BAK	Babble OVRL	Car SIG	Car BAK	Car OVRL	Station SIG	Station BAK	Station OVRL	train SIG	train BAK	train OVRL
	GSS	2.08	1.46	1.68	2.29	1.63	1.85	2.00	1.52	1.63	1.04	1.62	1.69
	mcra	2.24	1.19	1.84	2.47	1.31	2.04	2.35	1.23	1.92	2.35	1.33	1.96
PGSS	mcra2	2.20	1.20	1.82	2.40	1.30	1.99	2.29	1.24	1.89	2.33	1.33	1.94
	imcra	2.31	1.22	1.89	2.46	1.30	2.02	2.39	1.26	1.96	2.39	1.35	1.99
	martin	2.48	1.31	2.00	2.60	1.37	2.09	2.57	1.34	2.05	2.44	1.37	1.99
SNR= 5dB		Babble SIG	Babble BAK	Babble OVRL	Car SIG	Car BAK	Car OVRL	Station SIG	Station BAK	Station OVRL	train SIG	train BAK	train OVRL
	GSS	2.59	1.84	2.12	2.78	1.96	2.26	2.63	1.91	2.15	2.53	1.96	2.11
	mcra	2.85	1.46	2.36	3.05	1.59	2.53	3.03	1.55	2.51	2.87	1.59	2.41
PGSS	mcra2	2.75	1.44	2.29	2.96	1.56	2.46	2.89	1.53	2.42	2.78	1.56	2.34
	imcra	2.88	1.48	2.38	3.06	1.59	2.53	3.04	1.57	2.52	2.88	1.59	2.41
	martin	3.02	1.56	2.46	3.10	1.60	2.51	3.15	1.60	2.56	2.91	1.60	2.39

Table 2 lists the PESQ score obtained after processing, we observe that the PESQ score is consistent with the subjectively perceived trend of an improvement in speech quality with the proposed speech enhancement approach over that the spectral subtractive algorithm alone. This improvement is particularly significant in the case of car noise. Table 3 and table 4 list at different signal to noise ratio the subjective overall quality the OVRL measure that includes the naturalness of speech (SIG) and intrusiveness of background noise (BAK). We notice that the proposed auditory spectral

Table 4. SIG- BAK -OVRL for the proposed perceptual generalized spectral subtraction method (PGSS) compared to the subtractive algorithms at 10 and 15 dB

SNR=10dB		Babble			Car			Station			train		
		SIG	BAK	OVRL	SIG	BAK	OVRL	SIG	BAK	OVRL	SIG	BAK	OVRL
	GSS	2.99	2.16	2.49	3.09	2.23	2.57	3.01	2.20	2.49	2.93	2.21	2.45
	mcra	3.47	1.77	2.89	3.53	1.84	2.97	3.55	1.82	2.98	3.38	1.84	2.85
PGSS	mcra2	3.32	1.73	2.79	3.42	1.81	2.90	3.40	1.78	2.87	3.27	1.78	2.77
	imcra	3.49	1.79	2.91	3.55	1.84	2.98	3.56	1.84	2.99	3.41	1.82	2.86
	martin	3.55	1.82	2.92	3.60	1.84	2.95	3.63	1.85	2.98	3.43	1.82	2.82

SNR= 15dB		Babble			Car			Station			train		
		SIG	BAK	OVRL	SIG	BAK	OVRL	SIG	BAK	OVRL	SIG	BAK	OVRL
	GSS	3.33	2.41	2.79	3.41	2.44	2.83	3.28	2.39	2.73	3.29	2.44	2.76
	mcra	3.95	205	3.35	4.06	2.10	3.45	3.99	2.05	3.37	3.92	209	3.34
PGSS	mcra2	3.81	2.00	3.24	3.92	2.06	3.35	3.87	2.02	3.29	3.78	2.04	2.77
	imcra	3.98	2.05	3.36	4.09	2.11	3.46	4.01	2.06	3.39	3.95	2.09	3.36
	martin	4.00	2.06	3.35	4.06	2.07	3.38	4.07	2.07	3.38	3.92	2.05	3.27

attenuation using different continuous noise estimation algorithms performed significantly better than the classic subtractive attenuation. Lower distortion (higher BAK score) is observed in most condition. This demonstrates the performance of our approach to reduce the noticeable of the background noise. In the other hand, lower signal distortion (higher SIG score) is observed with the proposed approach in most condition with significant differences in 10dB for car noise. Also, we notice that incorporating continuous noise estimation in particularly the IMCRA estimation in the auditory spectral attenuation approach performed better than the spectral subtraction and the Ephraim and Malah rules in overall quality. This indicates that the proposed auditory spectral attenuation for speech enhancement is sensitive to the noise spectrum estimate. The results obtained show that the proposed speech enhancement method using different continuous noise estimation performed, in most condition, better than the classic spectral attenuation algorithms in terms of perceptual improvement, overall quality and low signal distortion. The auditory spectral analysis contributed significantly to the speech enhancement and to the improvement of the voice quality at different signal to noise ratio and practically for all the types of noise. Indeed the decomposition in filterbank and the continuous noise estimation given the best subjective results with regard to the subtractive noise reduction method based on uniform decomposing using Fourier transform and a simple approach to estimate the noise during the silent moment.

4   Conclusion

In this paper, we proposed a new speech enhancement method which consists in integrating perceptual proprieties of the human auditory system; it is based on decomposing the input signal in non-uniform sub-bands using an analysis/synthesis gammachirp filterbank that are manipulated in each frequency bin with the generalized spectral subtraction process. We noticed that the use of a frequency resolution according to the ERB scale, allowed obtaining, from the perceptive point of view and from the vocal quality, better results than those supplied by the classic spectral sub-traction in improving the quality and intelligibility of the enhanced speech signal.

References

1. Berouti, M., Schwartz, R., Makhoul, J.: Enhancement of speech corrupted by acoustic noise. In: Proc. Int. Conf. on Acoustics, Speech, Signal Processing, April 1979, pp. 208–211 (1979)
2. Ephraim, Y., Malah, D.: Speech enhancement using a minimum mean-square error short-time spectral amplitude estimator. IEEE transaction on acoustics speech and signal processing assp-33(2) (December 1985)
3. Tsoukalas, D.E., Mourjopoulos, J.N., Kokkinakis, G.: Speech enhancement based on audible noise suppression. IEEE Trans. Speech and Audio Processing 5, 497–514 (1997)
4. Virag, N.: Single channel speech enhancement based on masking properties of the human auditory system. IEEE Trans. Speech and Audio Processing 7, 126–137 (1999)
5. Hohmann, V.: Frequency analysis and synthesis using a Gammatone filterbank. Acta Acustica united with Acustica 88(3), 433–442 (2002)
6. ITU-T P.862, Perceptual evaluation of speech quality (PESQ), an objective method for end-to-end speech quality assessment of narrow-band telephone networks and speech codecs, International Telecommunication Union, Geneva (2001)
7. Loizou, P.: Speech Enhancement: Theory and Practice. CRC Press, Boca Raton
8. Martin, R.: Noise power spectral density estimation based on optimal smoothing and minimum statistics. IEEE Trans. on Speech and Audio Processing 9(5), 504–512 (2001)
9. Cohen, I., Berdugo, B.: Noise estimation by minima controlled recursive averaging for robust speech enhancement. IEEE Signal Proc. Letters 9(1), 12–15 (2002)
10. Cohen, I.: Noise Spectrum estimation in adverse environnements: improved minima controlled recursive averaging. IEEE Trans. Speech Audio Process. 11(5), 466–475 (2003)
11. Rangachari, S.,, P.: A noise estimation algorithm with rapid adaptation for highly non-stationary environments. In: IEEE Int. Conf. on Acoustics, Speech, signal processing, May 17-21, pp. I-305–I-308 (2004)
12. Irino, T., Unoki, M.: An analysis/synthesis auditory filterbank based on an IIR gammachrp filter. In: Greenberg, S., Slaney, M. (eds.) Computational models of Auditory Function. NATO ASI series, vol. 312. IOS Press, Amsterdam (2001)

Coding of Biosignals Using the Discrete Wavelet Decomposition

Ramon Reig-Bolaño[1], Pere Marti-Puig[1], Jordi Solé-Casals[1], Vladimir Zaiats[1], and Vicenç Parisi[2]

[1] Digital Technologies Group, University of Vic, C/ de la Laura 13, E-08500, Vic, Spain, EU
[2] Electronic Eng. Dep. Polith. Univ. of Catalonia,
{ramon.reig,pere.marti,jordi.sole,vladimir.zaiats}@uvic.cat,
Vicenc.Parisi@upc.edu

Abstract. Wavelet derived codification techniques are widespread used in image codifiers. The wavelet based compression methods are adequate for representing transients. In this paper we explore the use of the discrete wavelet transform analysis of biological signals in order to improve the data compression capability of data coders. The wavelet analysis provides a subband decomposition of any signal, and this enables a lossless or a lossy implementation with the same architecture. The signals could range from speech to sounds or music, but the approach is more orientated to other biosignals like medical signals EEG, ECG or discrete series. Experimental results based on wavelet coefficients quantification, show a lossless compression of 2:1 in all kind of signals, with a fidelity, measured using PSNR, from 79dB to 100dB, and lossy results preserving most of the signal waveform, with a compression ratio from 3:1 to 5:1, with a fidelity from 25dB to 35 dB.

Keywords: data compression, source coding, discrete wavelet transform.

1 Introduction

Data compression or source coding techniques try to use the minimum of bits/s to represent information of a source; they could be classified in lossless compression - totally reversible- and lossy compression, allowing better compression rates with some distortion on reconstructed signal. Lossless compression schemes usually exploit statistical redundancy and are reversible, so that the original data can be reconstructed [1]; while lossy data compression are usually guided by research on how people perceive the data, and accept some loss of data in order to achieve higher compression. In speech and music coding [2],[3], there are several standards like CELP (used in digital telephony), or the family of MP3 (MPEG 1 layer 3, for audio coding). There are two basic lossy compression schemes: in lossy predictive codecs (i.e. CELP), previous and/or subsequent decoded data are used to predict the current sound sample or image frame [4], the error between the predicted data and the real data, together with any extra information needed to reproduce the prediction, is then quantized and coded; by the other hand, in lossy transform codecs (i.e MP3), samples of picture or sound are taken, they are chopped into small segments, and they are transformed into a new basis space and quantized. In some systems the two

J. Solé-Casals and V. Zaiats (Eds.): NOLISP 2009, LNAI 5933, pp. 144–151, 2010.

techniques are combined, with transform codecs being used to compress the error signals generated by the predictive stage. In a second step, the resulting quantized values are then coded, using lossless coders (like run-length or entropy coders).

In this paper we explore the use of the discrete wavelet transform analysis of an arbitrary signal in order to improve the data compression capability of the first step. Wavelet analysis is widespread used in image codifiers [4],[5], for example in JPEG2000: using a 5/3 wavelet for lossless (reversible) compression and a 9/7 wavelet (Cohen-Daubechies-Feauveau biorthogonal wavelet [6]) for lossy (irreversible) compression.

The wavelet compression methods are adequate for representing transients [5], such as percussion sounds in audio, or high-frequency components in two-dimensional images, for example an image of stars on a night sky. This means that the transient elements of a data signal can be represented by a smaller amount of information than would be the case if some other transform, such as the more widespread discrete cosine transform or the discrete Fourier transform, had been used.

This paper investigates the application of discrete wavelet transforms, specifically the Cohen-Daubechies-Feauveau biorthogonal wavelet [6], for the compression of different kinds of unidimensional signals, like speech, sound, music or others like EEG [7], ECG or discrete series. Preliminary experimental results show a lossless compression of 2:1 in all kind of signals, with a fidelity -measured using PSNR (Peak to Signal Noise Ratio, see Annex)- from 79dB to 100dB; and lossy results preserving most of the signal waveform of about 5:1 to 3:1, with a fidelity from 25dB to 35 dB.

2 Discrete Wavelet Transform Analysis

The discrete wavelet transform analysis could be seen as a method for subband or multiresolution analysis of signals. The fast implementation of discrete wavelet transforms is done with a filter bank for the analysis of signals, and the inverse transform is done with a synthesis filter bank (Fig.1) [5]. The first application of this scheme was in speech processing as QMF (Quadrature Mirror Filter) [8] and [9].

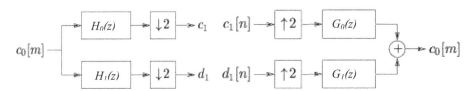

Fig. 1. Analysis filter bank, for Fast Discrete Wavelet Transform implementation. Approximation coefficients $c_1[n]$ are the Low-Pass part of $c_0[m]$, and Detail coefficients $d_1[n]$ correspond to the High-Pass part. And Synthesis Filter Bank, for Fast Discrete Inverse Wavelet Transform.

This decomposition could be iterated on the low pass component on successive stages, leading to an octave band filter bank [10], in Fig. 2, with frequencies responses of Fig. 3, if the decomposition has three levels.

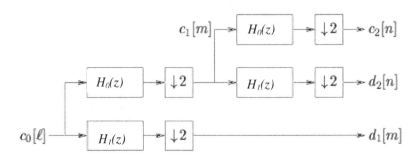

Fig. 2. A two level Analysis Filter bank implementation

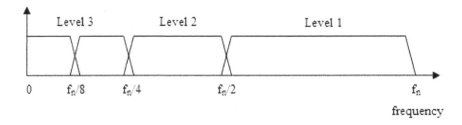

Fig. 3. Frequency response of Discrete Wavelet Analysis with 3 levels. Detail signal $d_1(m)$ is in the band from $f_n/2$ to f_n , $d_2(m)$ in the band from $f_n/4$ to $f_n/2$, $d_3(m)$ is from $f_n/8$ to $f_n/4$, and approximate signal $c_3(m)$ is from 0 to $f_n/8$.

In order to achieve perfect reconstruction the relation of Z-Transform of the filters like 9/7 Cohen-Daubechies-Feauveau biorthogonal wavelet [6], must accomplish,

$$H_1(z) \cdot G_1(z) + H_0(z) \cdot G_0(z) = 2$$
$$G_0(z) \cdot H_0(-z) + G_1(z) \cdot H_1(-z) = 0$$

$$(1)$$

3   Quantization of Wavelet Coefficients at All the Stages, Midtread vs. Midrise

Most of the compression methods based on transforms achieve their compression rate using fewer bits than the original ones with the codification of transformed coefficients [4]. Depending on the number of bits assigned we define the number of quantification steps. In the case of 3 bits for quantification, the quantification could be done with a midtread quantizer Fig. 4 (a), using 7 levels, or with a midrise quantizer Fig. 4 (b), using 8 levels.

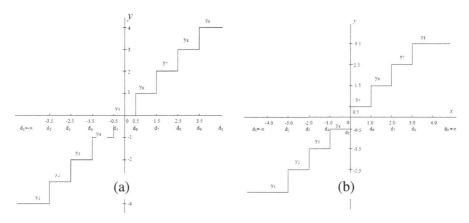

Fig. 4. 3 bits midtread uniform quantizer (a) with 7 levels and 3 bits midrise uniform quantizer (b) with 8 levels of quantification

4   Proposed Method

We propose a data compressor following the scheme of Fig. 5, based on a Discrete Wavelet Analysis of the signal with the filter bank of the Cohen-Daubechies-Feauveau biorthogonal wavelet [6]. In this first experiment the system will have two entry parameters: the first one will be the number of stages or levels of decomposition; and the second parameter will be the numbers of bits used to quantify the components at each stage.

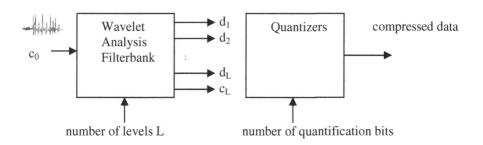

Fig. 5. Block diagram of the proposed method for the codification

In this experiment to recover the signal we will apply a Wavelet Synthesis Filterbank on compressed data to get an approximation of the original signal. To complete the codification we should use a lossless compression on the compressed data (like run-length-coder or Lempel-Ziv [1]); this step however is out of the scope of this paper.

5 Experiments

To investigate the performance of the system we have done several experiments. We have used a voice signal (16 bits/sample, Fs=11025 Hz), a music signal (16 bits/sample and Fs=44100 Hz), and an EEG channel (16 bits/sample, Fs=250 Hz).

In the case of the voice signal (16 bits/sample, Fs=11025 Hz), the 3 stage DWA (Discrete Wavelet Analysis), leads to the signals of Fig. 6 (left). If we apply a quantification of the detail components with 3 bits, using a 7 level midtread quantifier, we get Fig. 6 (right), with a compression of 3.2:1 respect to the original data of 16 bits.

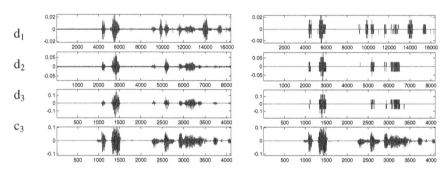

Fig. 6. Discrete Wavelet Analisis with 3 stages (left), and Quantified Discrete Wavelet Analysis with 3 bits (7 levels) (3.2:1 compression respect original signal 16 bits)

If we use this compressed data (Fig. 6) (3 bits with 3.2:1 compression) to recon-struct with the Synthesis Filter Bank, we obtain an approximation of original signal with a PSNR= 29 dB at Fig. 7. The voice is intelligible but rather noisy, compared with perfect reconstruction PSNR=100 dB, attained by the same system with 8 bits (2:1 compression).

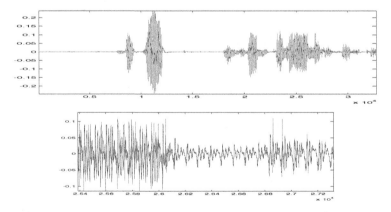

Fig. 7. Signal reconstruction (red), original voice signal (blue). Magnification of a nosy segment, PSNR=29 dB. Lossless reconstruction PSNR=100 dB.

In the case of the music signal (16 bits/sample, Fs=44100 Hz), the 3 stage DWA (Discrete Wavelet Analysis), leads to the signals of Fig. 8 (left). If we apply a quantification of the detail components with 3 bits, using a 7 level midtread quantifier, we get Fig. 8 (right), with a compression of 3.2:1 respect to the original data of 16 bits).

d_1
d_2
d_3
c_3

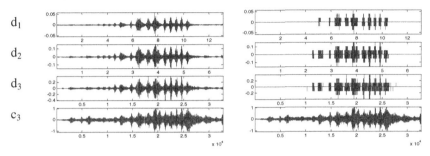

Fig. 8. Discrete Wavelet Analisis with 3 stages (left), and Quantified Discrete Wavelet Analysis with 3 bits (7 levels) (3.2:1 compression respect original signal 16 bits)

If we use this compressed data (Fig. 8) (3 bits with 3.2:1 compression) to reconstruct with the Synthesis Filter Bank, we obtain an approximation of original signal with a PSNR=32 dB at Fig. 9. The music is rather noisy, compared with perfect reconstruction PSNR= 97 dB, attained by the same system with 8 bits (2:1 compression).

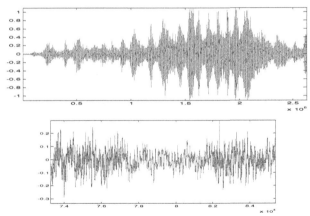

Fig. 9. Signal reconstruction (red), original voice signal (blue). PSNR=32 dB. Lossless reconstruction with music PSNR=97 dB.

In the case of the EEG signal (16 bits/sample, Fs=250 Hz), the 3 stage DWA (Discrete Wavelet Analysis), leads to the signals of Fig. 10 (left). If we apply a quantification of the detail components with 3 bits, using a 7 level midtread quantifier, we get Fig. 10 (right), with a compression of 3.2:1 respect to the original data of 16 bits.

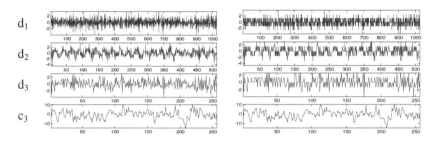

Fig. 10. Discrete Wavelet Analisis with 3 stages (left), and Quantified Discrete Wavelet Analysis with 3 bits (7 levels) (3.2:1 compression respect original signal 16 bits)

If we use this compressed data (Fig. 10) (3 bits with 3.2:1 compression) to reconstruct with the Synthesis Filter Bank, and use a midtread we obtain an approximation of original signal with a PSNR=27.48 dB at Fig. 11. The signal is quite similar to original, but with some distortion compared with perfect reconstruction PSNR= 77.6 dB, attained by the same system with 8 bits (2:1 compression).

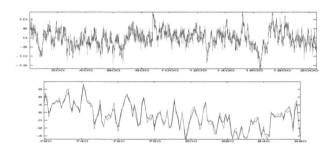

Fig. 11. Signal reconstruction (red), original EEG signal (blue). Magnification from sample 720 to 860. PSNR=27.48 dB. Lossless reconstruction with PSNR=77.6 dB.

6 Conclusions

In this paper, the use of discrete Wavelet Transform Analysis to improve a data compression codifier is explored. The preliminary results are highly positive; we reach a 2:1 of compression rate on the lossless codifiers. The compression is obtained with the quantification of detail coefficients of Discrete Wavelet Analysis with 8 bits. Quantitative measures give better PSNR (PSNR=100 dB) with our approach than with a direct quantification of original signals with 8 bits (PSNR=50 dB), the voice signal becomes noisy; the gain on PSNR with the same rate of compression is about 50 dB.

On the other hand, with the lossy compression, we reach a scalable compression rate that ranges to from 3.2:1 to 5:1 of compression rate with PSNR results from 25 dB to 35 dB. These results are relevant and promising for the compression of natural signals, like EEG channels or ECG signals, with a method that preserves transient and other relevant characteristics of the waveform.

Future work will be in several directions: explore the use of frames of signals, instead of using global signals; measure the impact of different quantifiers on the fidelity of the reconstructed signals; and finally a systematic study to extend these preliminary results to a wide range of measures and signals, including the effect of noise on coded data.

Annex *PSNR*

To measure quantitatively the fidelity of a reconstructed signal we use the *PSNR* (*Peak to Signal Noise Ratio*), we have an N samples original signal c_0 , and a reconstructed signal $\hat{c}_0$ then the *PSNR*,

$$RMSE = \sqrt{\frac{\sum_{i=1}^{N}(\hat{c}_0(i) - c_0(i))^2}{N}}, \tag{2}$$

$$PSNR = \frac{\max(|c_0|)}{RMSE}$$

Acknowledgments. This work has been supported by the University of Vic under the grant R0904.

References

1. Ziv, J., Lempel, A.: Compression of individual sequences via variable-rate coding. IEEE Transactions on Information Theory 24(5), 530–536 (1978)
2. Spanias, A.S.: Speech coding: a tutorial review. Proceedings of the IEEE 28(10), 1541–1582 (1994)
3. Gersho, A.: Advances in speech and audio compression. Proceedings of the IEEE 82(6), 900–918 (1994)
4. Jain, A.K.: Image data compression: A review. Proceedings of the IEEE 69(3), 349–389 (1981)
5. Mallat, S.: A wavelet tour of signal processing, 2nd edn. Academic Press, London (1999)
6. Cohen, A., Daubechies, I., Feauveau, J.C.: Biorthogonal bases of compactly supported wavelets. Commun. on Pure and Appl. Math. 45, 485–560 (1992)
7. Antoniol, G., Tonella, P.: EEG Data Compression Techniques. IEEE Trans. on Biomedical Engineering 44(2) (February 1997)
8. Esteban, D., Galand, C.: Application of quadrature mirror filters to split band voice coding schemes. In: Proc. of 1977 ICASSP, May 1977, pp. 191–195 (1977)
9. Smith, M.J., Barnwell, T.P.: Exact reconstruction for tree structured subband coders. IEEE Trans. Acoust., Speech, and Signal Proc. 34(3), 431–441 (1986)
10. Proakis, J.G., Manolakis, D.: Digital Signal Processing: Principles, Algorithms and Applications, 3rd edn. Prentice-Hall, Englewood Cliffs (1996)

Reducing Features from Pejibaye Palm DNA Marker for an Efficient Classification

Carlos M. Travieso, Jesús B. Alonso, and Miguel A. Ferrer

Signals and Communications Department, Technological Centre for Innovation in Communications, Campus Universitario de Tafira, s/n, Ed. Telecomunicación, Pab. B – D-111
University of Las Palmas de Gran Canaria
E-35017, Las Palmas de Gran Canaria, Spain
ctravieso@dsc.ulpgc.es

Abstract. This present work presents different feature reduction methods, applied to Deoxyribonucleic Acid (DNA) marker, and in order to identify a success of 100% based on Discriminate Common Vectors (DCV), Principal Component Analysis (PCA), and Independent Component Analysis (ICA) using as classifiers Support Vector Machines (SVM) and Artificial Neural Networks. In particular, the biochemical parameterization has 89 Random Amplified polymorphic DNA (RADPS) markers of Pejibaye palm landraces, and it has been reduced from 89 to a 3 characteristics, for the best method using ICA. The interest of this application is due to feature reduction and therefore, the reduction of computational load time versus the use of all features. This method allows having a faster supervised classification system for the process of the plant certification with origin denomination. Therefore, this system can be transferred to voice applications in order to reduce load time, keeping or improving the success rates.

Keywords: Feature reduction, Discriminative Common Vector, Independent Component Analysis, Principal Component Analysis, supervised classification, DNA markers.

1 Introduction

The pejibaye palm belongs to the monocotyledons, family of the Arecaceae, tribe of the cocoids, sub tribe Bactridinae and Bactris genus. This palm is the only domesticated one of the neotropic and produce: fruit, wood, and the most common and know heart-of-palm "palmito" present on international markets. This palm presents a large variety of morphology genus and large distribution over Central and South America. In this paper, authors present different methods for reducing the number of Pejibaye palm DNA markers. Feature reduction for pattern recognition is an important tool due to the great number of information fonts and/or parameters from patterns. Try concentrating the information or to create techniques for this aim, it is a point very requested for complex system.

J. Solé-Casals and V. Zaiats (Eds.): NOLISP 2009, LNAI 5933, pp. 152–162, 2010.

In order to certify this method, Support Vector Machines (SVM) [1] has been used and Artificial Neural Networks [2] as classifiers. The goal of this method is to get a smaller vector than the original vector with the same level of discrimination.

This kind of palm presents a large variety of morphology genus and large distribution over Central and South America (Mesoamerican area). Since last century, due to the crop origin controversy [3], [4] till now unsolved, mayor concern has been to identify biologically domestic races, and the research has been aimed to obtain genetic improvement and preservation instead of varieties identification. Till now, there is not known on scientific literature an automatic Pejibaye identification system. It has an interesting point of economic view because different landraces (varieties) provide one or another product, but firstly, it has to be identified its origin denomination. It shows an evident interest to certify correctly each one of different varieties.

In the figure 1 is shown the general system for supervised classification. It has been implemented for all feature vector and applying the feature reduction based on Discriminative Common Vector (DCV) [5], PCA [6], and ICA [7]. This method has been evaluated with a database of DNA markers, but it can be applied on voice applications or other applications on N-dimension. In this application, this method has reduced on three times the number of parameters, keeping the level of discrimination after the feature reduction.

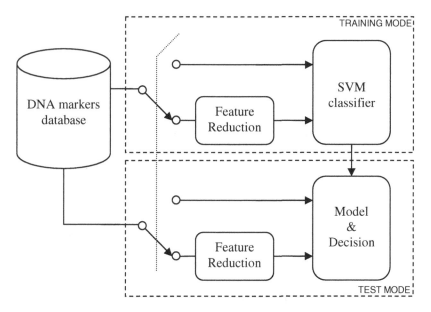

Fig. 1. System proposed with or without feature reduction, and working on two modes of the supervised classification

On this study we have achieved three important results. Firstly, the validation of RAPDS (Random Amplified polymorphic DNA) traces analysis technique with the system proposed. Therefore it is an important item for the Biochemistry because those RAPDS markers there were not validated for Pejibaye palm. Secondly, a substantial features reduction, allowing a better response time of system. And finally, a 100% correct identification of each palm variety.

2   Pejibaye Palm Database

For this present work, it has been considered six landraces pejibaye palms: Utilitis (Costa Rica), Tuira (Panama), Putumayo (Colombia), Yurimagua (Peru), Tembé (Bolivia) and Pará (Brazil). Selected criterion considered for races has been proponed by Clement and Mora-Urpi [3], [4] and [8]. Each class has 13 samples with 89 RAPDS markers per sample. This database is so much reduced, but it is very expensive the implementation of Biochemistry method. In future works and after achieve funds we hope to increase this database.

Deoxyribonucleic acid (DNA) is a long polymer of nucleotides, with a backbone made of sugars and phosphate groups joined by ester bonds. Attached to each sugar is one of four types of bases molecules and, it is the sequence of these four bases along the backbone that encodes information. This code is read by copying chains of DNA into the related nucleic acid RNA.

Raw DNA analysis is a very expensive and time consuming technique but, the interest of such analysis is based on the fact that it is used on decision making, handled and preservation of genetic resources, taxonomy and systematic molecular studies. Several techniques have been developed in order to diminish this description extension. RAPDS trace analysis (Random Amplified polymorphic DNA) is similar to a fingerprint technique because it identifies species, but based on PCR (Polymerase Chain Reaction) [9], [10], [11], [12] and [13].

This study was realized over each individual's genetic material, with 89 OPC primers (from the Operon Company) obtaining information variables with clear and well defined fragments, after multiples reaction amplifications. It was obtained for each Pejibaye Palm, a vector with 89 binary parameters. That is to say, markers and individuals produced a binary matrix, indicating enough presence of a particular RAPDS primer, from the six different Pejibaye landraces considered.

3   Feature Reduction

In this present work, three feature reduction have been used, Discriminate Common Vector (DCV), Principal Component Analysis (PCA) and Independent Component Analysis (ICA).

3.1   Discriminative Common Vector

It has been implemented as parameterization system, the Discriminative Common Vectors (DCV) technique [5] and [14]. It extracts the common characteristics of each class eliminating the difference between samples in each class. A common vector is obtained eliminating all the features that are in the direction of eigenvectors corresponding to the non-null eigenvalues of the scatter matrix of its own class. After its calculation, the common vectors are used for the identification. The implementation of the algorithm in order to do DCV method is known as the computation of DCV using the space of S_w range [14].

The first step is the calculation of S_w range. Therefore, it is calculated the non-null eigenvalues and its corresponding eigenvectors of $S_w = AA^T$ by the use of the matrix AA^T, where A is given in the following equation;

$$A = [x_1^1 - \mu_1, \ldots x_N^1 - \mu_1, x_1^2 - \mu_2, \ldots x_N^C - \mu_C] \tag{1}$$

where x_i^j is the characteristic i of the class j, and μ is its mean. Make $Q=[\alpha_1, \ldots, \alpha_r]$, where α_i is the set of orthonormal eigenvectors corresponding to the nonzero eigenvalues of S_w, and r is the dimension of S_w. And finally, it is built with Q, the projection matrix $P=QQ^T$.

The second step is to choose any vector of the sample and apply on the null space of S_w in order to get the common vectors;

$$x_{com}^i = x_m^i + Q\bar{Q}x_m^i \tag{2}$$

where $m = 1, \ldots, N$ classes and $i = 1, \ldots, C$ characteristics.

The following step is to calculate the Principal components with the eigenvectors w_k of S_{com}, which correspond to the non-null eigenvalues,

$$J(W_{opt}) = \arg \max_W |W^T S_{com} W| \tag{3}$$

using the matrix $A^T_{com}A_{com}$, where $S_{com}=A^T_{com}A_{com}$ y A_{com} are given in the equation 4;

$$A_{com} = [x_{com}^1 - \mu_{com} \ldots x_{com}^C - \mu_{com}] \tag{4}$$

In the step 4, C-1 common vectors correspond to the non-null eigenvalues. Use those vectors in order to build the projection matrix $W=[w_1, \ldots, w_{C-1}]$, it will be useful to get the characteristic vectors with the equations 3 and 4,

The fifth step is to choose a characteristic vector per each class and apply it to W, the result will be a classification vector, that is, its proximity will belong to the class or not.

Finally in the step 6, in order to apply the final characteristic vector, there will be to unite the first transformation P with the second W on an only projection matrix W, which will use the test vector in the characteristic space.

$$\Omega_{test} = W^T x_{test} \tag{5}$$

3.2   Principal Component Analysis

Principal Components Analysis (PCA) is a way of identifying patterns in data, and expressing the data in such a way as to highlight their similarities and differences [6]. Since patterns in data can be hard to find in data of high dimension, where the luxury of graphical representation is not available, PCA is a powerful tool for analyzing data. The other main advantage of PCA is that once you have found these patterns in the data, and you compress the data, i.e. by reducing the number of dimensions, without much loss of information.

PCA is an orthogonal linear transformation that transforms the data to a new coordinate system such that the greatest variance by any projection of the data comes to lie on the first coordinate (called the first principal component), the second greatest variance on the second coordinate, and so on. PCA is theoretically the optimum transform for a given data in least square terms.

In PCA, the basis vectors are obtained by solving the algebraic eigenvalue problem $\mathbf{R}^T(\mathbf{XX}^T)\mathbf{R} = \Lambda$ where $\mathbf{X}$ is a data matrix whose columns are centered samples, $\mathbf{R}$ is a matrix of eigenvectors, and Λ is the corresponding diagonal matrix of eigenvalues. The projection of data $\mathbf{C}_n = \mathbf{R}_n^T\mathbf{X}$, from the original p dimensional space to a subspace spanned by n principal eigenvectors is optimal in the mean squared error sense.

3.3 Independent Component Analysis

The main objective of Independent Component Analysis (ICA) [7] is to obtain, from a number of observations, the different signals that compose these observations. This objective can be reached using two different techniques, the spatial and the statistical. The first one is based on an input signal and depends on the position and separation of them. On the other hand, the statistical separation supposes that the signals are statistically independent, that they are mixed in a linear way.

This technique is studied in this work. And it is used in different fields such as real life applications [15] and [16], natural language processing [17], bioinformatics, image processing [18], etc.

ICA comes from PCA. PCA is used in a wide range of scopes such as face recognition or image compression, being a very common technique to find patterns in high dimension data.

ICA can study two different problems; the first one is when the mixtures are linear, while the second one, due to when the mixtures are convolutive and they are not totally independent because the signal has been through dynamic environments. It is the "Cocktail party problem".

Depending on the mixtures, there are several methods to solve the Blind Source Separation (BSS) problem. The first case can be seen as a simplification of the second one.

The blind source separation based on ICA is also divided into three groups; the first one are those methods that works in the time domain, the second are those who works in the frequency domain and the last group are those methods that combine frequency and time domain methods. For our case, we will work with spatial domain.

4   Classification System

In this work two different classification systems have been used in order to offer an independent view. Therefore, the features or parameter will be evaluated using Support Vector Machines and Artificial Neural Networks.

4.1 Support Vector Machine

On this work we have used a Support Vector Machines (SVM) in order to evaluate and analyze performance and behavior to DNA parameterization of Pejibaye palms.

The classification model based on SVM is a bi-classes system; only it is able discriminate between two classes. SVM is based on the concept of decision planes. That decision plane is defined as the separator plane between a set of samples of both classes according to Karush-Kuhn-Tucker's conditions (KKT). From a set of geometric properties, the model computes the cited plane [1] and [19]. Particularly, we have used an implementation of Vapnik`s Support Vector Machine known as SVM light [1] and [19], which is a fast optimization algorithm for pattern recognition, regression problem, and learning retrieval functions from unobtrusive feedback to propose a ranking function. The optimization algorithms used in SVM light are described in [1] and [19]. In order to obtain no-lineal SVM are used the mathematical functions called kernels. SVM transforms the features to a space of very high dimension by that kernel function.

To apply this method to multi-classes problems, the technique of one-versus-all is applied. It consists of the building of M binary classifiers; each classifier is trained for discriminating a class versus M-1 rest of classes.

The classification system has been used as supervised classification, and therefore, it consists of two modes, training and test modes. The first mode needs to be trained in order to build a model, subsequently used on the test mode (see figure 1). For another mode, the system is able to decide with an unknown sample and using the previous model, which is class that belong to the sample analysed.

4.2 Artificial Neural Network

An Artificial Neural Network (ANN) can be defined as a distributed structure of parallel processing, formed by artificial neurons, interconnected by a great number of connections (synapses) (see Fig. 2) [2].

Feed-Forward networks consist of layers of neurons where the exit of a neuron of a layer feeds all the neurons on the following layer. The fundamental aspect of this structure is that feedback unions do not exist. The so called Multilayer Perceptron (MLP) is a type of Feed-forward ANN, where the threshold function is a nonlinear

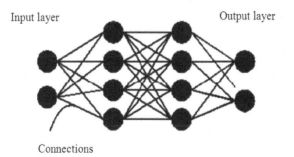

Fig. 2. General Neural Network structure

but differentiable function. This nonlinear differentiable function is necessary to assure that the gradient can be calculated, in order to find the good values for all the parameters of the network.

For the classification system we have used a 2-layer feed-forward perceptron trained by means of conjugated gradient descent algorithm [2], doing tests from 20 to 150 neurons in the hidden layer and hyperbolic tangent *tanh(.)* as a nonlinear function for these units. The number of input neurons fits in with the number of components (from 1 to 89 markers), and the number of output neurons with the number of kinds of Pejibaye (6 in our application).

5   Experiments and Results

It has been developed a supervised identification system, using the crossing validation method. Experiments have been done 10 times, and therefore, the success rate is shown with mean and variance (see tables 1 and 2). This work presents two experiments, first one is the evaluation of the feature reduction and the second one is the study of the load time.

On the one hand, we are showing the success with the use of feature reduction methods and without it. Using in all cases as classifiers, ANN and SVM with lineal and RBF kernels, where γ is applied on the exponential function of RBF kernel. Success rates applied with ANN are not shown due to rates were lower than success rates based on SVM, lower than 10%.

Without the feature reduction, we have used between 8% and 23% (from 1 to 3 samples/class) from our database in order to do training mode, and the rest of them (from 92% - 12 samples/class to 77% - 10 samples/class) for test mode. In the table 1 is shown success rates.

In the table 2 is shown the results applying DCV method. The samples of training mode were between 23% and 54% from our database, and the rest of samples for test mode. Besides, it can be seen in table 2 the different successes with respect to the size of feature reduced vector and the threshold of non-null eigenvalue, where that number is equivalent to null.

In the table 3 is shown the results applying PCA method. The samples of training mode were between 8% and 60% from our database, and the rest of samples for test mode. The feature reduction was from 89 to 4 characteristics. With a number of characteristics inferior to 4, it was not possible to achieve 100% of success rate.

Table 1. Results using all original DNA markers with the SVM classifier

Training Percentage	Success Rates	Type of kernel	γ
8%	98.89% ± 0.34	Lineal	---
15%	99.85% ± 0.23	Lineal	---
23%	100% ± 0	Lineal	---
8%	99.02% ± 0.88	RBF	5×10^{-2}
15%	100% ± 0	RBF	8×10^{-2}
23%	100% ± 0	RBF	8×10^{-2}

Finally, in table 3 is observed the results for ICA method. The samples of training mode were between 8% and 30% from our database, and the rest of samples for test mode; where the feature reduction was from 89 to 3 characteristics. Under this number of characteristics was not possible to achieve 100% of success rate.

Table 2. Results with DCV reduction from original DNA markers using the SVM classifier

Feature reduced - Training Percentage	Non-null Eigenvalue	Success Rates for RBF kernel	γ	Success Rates for lineal kernel
32 -23%	10^{-3}	92.67% ±1.48	1×10^{-1}	91.16% ±2.23
32 -31%	10^{-2}	95.27% ±1.25	1×10^{-1}	94.72% ±1.63
32 -38%	10^{-2}	98.41% ±1.22	5×10^{-3}	97.22% ±1.79
32 -46%	10^{-5}	98.14% ±1.75	5×10^{-3}	96.29% ±1.43
32 -54%	10^{-3}	98.88% ±1.72	1×10^{-1}	96.66% ±3.65
36 -23%	10^{-4}	98.73% ±1.14	1×10^{-2}	98.48% ±0.95
36 -31%	10^{-3}	98.05% ±2.21	1×10^{-1}	98.05% ±2.21
36 -38%	10^{-3}	99.38% 0.95	3×10^{-2}	98.77% ±1.54
36 -46%	10^{-1}	99.07% ±1.54	1×10^{-1}	98.61% ±1.52
36 -54%	10^{-5}	100% ± 0	8×10^{-2}	99.07% ±1.43
40 -23%	10^{-2}	98.33% ±1.05	1×10^{-1}	98.05% ±1.25
40 -31%	10^{-3}	99.44% ±0.86	5×10^{-2}	98.41% ±1.94
40 -38%	10^{-3}	99.69% ± 0.75	5×10^{-2}	99.07% ±1.43
40 -46%	10^{-3}	99.60% ±0.97	1×10^{-1}	99.38% ± 0.95
40 -54%	10^{-5}	100% ± 0	5×10^{-2}	100% ±0

Table 3. Results with PCA reduction from original DNA markers using the SVM classifier

Training Percentage	Success Rates	Type of kernel	γ
8%	92.42% ± 3.42	Lineal	---
15%	94.04% ± 3.23	Lineal	---
23%	96.50% ± 2.66	Lineal	---
30%	96.85% ± 2.15	Lineal	---
38%	96.04% ± 2.49	Lineal	---
45%	97.62% ± 1.94	Lineal	---
53%	99.12% ± 0.68	Lineal	---
60%	99.39% ± 0.57	Lineal	---
8%	93.94% ± 3.11	RBF	4×10^{-3}
15%	96.06% ± 2.28	RBF	3×10^{-3}
23%	97.17% ± 1.93	RBF	2×10^{-3}
30%	98.33% ± 1.37	RBF	1×10^{-4}
38%	98.75% ± 1.64	RBF	6×10^{-4}
45%	99.29% ± 1.15	RBF	1×10^{-3}
53%	99.66% ± 0.48	RBF	2×10^{-3}
60%	100% ± 0	RBF	5×10^{-3}

Table 4. Results with ICA reduction from original DNA markers using the SVM classifier

Training Percentage	Success Rates	Type of kernel	γ
8%	79.31% ± 7.30	Lineal	---
15%	81.06% ± 6.67	Lineal	---
23%	82.17% ± 3.24	Lineal	---
30%	99.81% ± 0.59	Lineal	---
8%	99.31% ± 1.35	RBF	1×10^{-2}
15%	99.39% ± 1.06	RBF	5×10^{-1}
23%	99.67% ± 0.70	RBF	5×10^{-1}
30%	100% ± 0	RBF	5×10^{-2}

In the tables 1 2, 3 and 4, it can be observed that it is achieved 100% for both case (with and without reduction), and therefore, this system can be used to validate a Pejibaye palm. Through in the feature reduction proposed method and for all cases, we need more samples for the training mode, in particular, and for the best case, 30% versus 23% with the full feature vector. But the advantage applying ICA reduction, it is that we have reduced 29 times the size of feature vector. Other methods need more training samples and more reduced characteristic in order to achieve a success of 100%.

This reduction method (and the others) can be applied to voice technologies, in particular on voice pathologies, because it have evaluated with a database with similar conditions, given a good result. Our database has few classes (types of pathologies), it is on 1-dimension, and too, the number of parameters is similar. In future works, we will transfer this method to the cited application in order to reduce the size of feature vector and the computational time, keeping the level of discrimination.

Successes rates were obtained with the proposal classification system (SVM with RBF kernel) using ICA as the reduction system (see figure 1). Characteristics reduction was applied to isolate the minimum components with the maximum threshold. In particular, our system reduced the number of required components from 89 to 3 to achieve a 100% success rate, with a resultant compression rate of 29,67:1. In addition, two kinds of classifiers were used, and a 100% success rate was achieved using 2 samples and training a SVM with RBF kernel (see table 4). Four training samples (30% from our database) were needed for our threshold (100%) on the 10 experiments.

Table 5. Computational load for systems proposed in milliseconds

Feature vector	Training Mode + reduction algorithm	Test Mode + reduction algorithm
Original	7012 ms + 0 ms	> 0.1 ms per sample + 0 ms
DCV	2787 ms + 14 ms	> 0.1 ms per sample + 12 ms
PCA	2124 ms + 43 ms	> 1 ms per sample + 39 ms
ICA	2542 ms + 82 ms	> 1 ms per sample + 80 ms

On the other hand, it has been checked the load time. The computational time is the average of 10 times of each experiment. It was calculated when a success rate was 100% in each case.

In table 5 is shown times for the all cases. For training mode, the reduced feature vector time is lower than original vector time; therefore, it is a reduction of time for all the cases, approximately three times. For test mode and all the cases, the time is very low and it is considered as real time.

6 Conclusions

In this paper we present a robust and efficient validation system for reducing features and doing an automatic identification of RAPDS makers of Pejibaye landraces, based on Independent Component Analysis and using a SVM with RBF kernel as classifier. We have got to reduce 29.67 times the number of features, and therefore, we have reduced the computational load time in this system for the training mode, approximately three times, and the system keeps real time for test mode. Moreover, it is kept the level of discrimination of system after to apply ICA method, and the rest of feature reduction proposed methods, PCA and DCV.

In future work, these methods will be applied to voice applications; in particular, voice pathologies detection, emotion detection for speech, and/or speaker recognition, and others.

Acknowledgments. This work was supported by "Programa José Castillejo 2008" from the Spanish Government. This research was made in University of Costa Rica.

References

1. Platt, J.C., Cristianini, N., Shawe-Taylor, J.: Large margin DAGs for multi-class classification. In: Advances in Neural Information Processing Systems, vol. 12, pp. 547–553. MIT Press, Cambridge (2000)
2. Bishop, C.M.: Neural Networks for Pattern Recognition. Oxford University Press, Oxford (1995)
3. Clement, C.R., Aguiar, J., Arkcoll, D.B., Firmino, J., Leandro, R.: Pupunha brava (Bactris dahlgreniana Glassman): progenitora da pupunha (Bactris gasipaes H.B.K.). Boletim do Museu Paraense Emilio Goeldi 5(1), 39–55 (1989)
4. Mora-Urpí, J., Clement, C., Patiño, V.: Diversidad Genética en Pejibaye: I. Razas e Híbridos. In: IV Congreso Internacional sobre Biología, Agronomía e Industrialización del Pijuayo, University of Costa Rica, pp. 11–20 (1993)
5. Cevikalp, H., Neamtu, M., Barkana, A.: IEEE Transactions on Systems, Man, and Cybernetics, Part B 37(4), 937–951 (2007),
   http://ieeexplore.ieee.org/xpls/abs_all.jsp?arnumber=4267852
6. Jolliffe, I.T.: Principal Component Analysis, 2nd edn. Springer Series in Statistics. Springer, NY (2002)
7. Hyvärinen, A., Karhunen, J., Oja, E.: Independent Component Analysis. John Wiley & Sons, New York (2001)

8. Mora-Urpí, J.: Arroyo. C.: Sobre origen y diversidad en pejibaye: Serie Técnica Pejibaye (Guilielma). Boletín Informativo. Editorial University of Costa Rica 5(1), 18–25 (1996)
9. Porebski, S., Grant, L., Baun, B.: Modification of a CTAB DNA extraction protocol for plants containing high polysaccharide and polyphenol components. Plant Molecular Biology Reporter, 15, pp. 8–15 (1997)
10. Ravishankar, K.V., Anand, L., Dinesh, M.R.: Assessment of genetic relatedness among mango cultivars of India using RAPD markers. Journal of Horticultural Sci. & Biotechnology 75, 19–201 (2000)
11. Williams, J.G.K.: DNA polymorphisms amplified by arbitrary oligonucleotide primers are useful as genetics markers. Nucleic Acids Research 18, 6531–6535 (1990)
12. Dellaporta, S.L., Wood, J., Hick, J.B.: Plant DNA minipreparation: Version II. Plant Mol. Biol. Rep. 1, 19–21 (1983)
13. Ferrer, M., Eguiarte, L.E., Montana, C.: Genetic structure and outcrossing rates in Flourensia cernua (Asteraceae) growing at different densities in the South-western Chihuahuan Desert. Annals of Botany 94, 419–426 (2004)
14. Cevikalp, H., Neamtu, M., Wilkes, M.: IEEE Transactions on Neural Networks 17(6), 1832–1838 (2006),
   http://eecs.vanderbilt.edu/CIS/pubs/kernel_dcv.pdf
15. Saruwatari, H., Sawai, K., Lee, A., Kawamura, T., Sakata, M., Shikano, K.: Speech enhancement and recognition in car environment using Blind Source Separation and subband elimination processing. In: Proceedings International Workshop on Independent Component Analysis and Signal Separation, pp. 367–372 (2003)
16. Saruwatari, H., Kawamura, T., Shikano, K.: Blind Source Separation for speech based on fast convergence algorithm with ICA and beamforming. In: Proceedings in EUROSPEECH, pp. 2603–2606 (2001)
17. Murata, N., Ikeda, S., Ziehe, A.: An approach to blind source separation based on temporal structure of speech signals. Neurocomputational 41, 1–24 (2001)
18. Cichocki, A., Amari, S.I.: Adaptive Blind Signal and Image Processing. John Wiley & Sons, Chichester (2002)
19. Joachims, T.: SVM_light Support Vector Machine, Department of Computer Science, Cornell University, Version: 6.02, http://svmlight.joachims.org/ (Last visit: March 2009)

Mathematical Morphology Preprocessing to Mitigate AWGN Effects: Improving Pitch Tracking Performance in Hard Noise Conditions

Pere Martí-Puig, Jordi Solé-Casals, Ramon Reig-Bolaño, and Vladimir Zaiats

Digital Technologies Group, University of Vic, Sagrada Família 7,
08500 Vic, Spain
{pere.marti,jordi.sole,ramon.reig,vladimir.zaiats}@uvic.cat

Abstract. In this paper we show how a nonlinear preprocessing of speech signal -with high noise- based on morphological filters improves the performance of robust algorithms for pitch tracking (RAPT). This result happens for a very simple morphological filter. More sophisticated ones could even improve such results. Mathematical morphology is widely used in image processing in where it has found a great amount of applications. Almost all its formulations derived in the two-dimensional framework are easily reformulated to be adapted to one-dimensional context.

Keywords: Robust Pitch Tracking, Mathematical Morphology, Nonlinear Speech Preprocessing.

1 Introduction

Pitch is a very important parameter in speech processing applications, such as speech analyzing, coding, recognition or speaker verification. Pith tracking also becomes relevant for the automatic recognition of emotions in spoken dialogues. Affective activity causes physiological variations reflected in the vocal mechanism and causes further speech variation being the pitch the most relevant acoustic parameter for the detection of emotions (Mozziconacci and Hermes, 1998, Juang and Furui, 2000, Petrushin, 2000 and Kang et al., 2000) [1-4]. For example, aroused emotions (such as fright and elation) are correlated with relatively high pitch, while relaxed emotions (such as tedium and sorrow) are correlated with relatively low pitch.

Pitch detection techniques are of interest whenever a single quasi-periodic sound source is to be studied or modeled [5][6]. Pitch detection algorithms can be divided into methods which operate in the time domain, frequency domain, or both. One group of pitch detection methods uses the detection and timing of some time domain feature. Other time domain methods use autocorrelation functions or some kind of difference of norms to detect similarity between the waveform and its time delayed version. Another family of methods operates in the frequency domain with the purpose of locating peaks. Other methods use combinations of time and frequency domain techniques to detect pitch. Frequency domain methods need the signal to be frequency transformed, and then the frequency domain representation is inspected for

J. Solé-Casals and V. Zaiats (Eds.): NOLISP 2009, LNAI 5933, pp. 163–170, 2010.
© Springer-Verlag Berlin Heidelberg 2010

the first harmonic, the greatest common divisor of all harmonics. Windowing of the signal is recommended to avoid spectral spreading, and depending on the type of window, a minimum number of periods of the signal must be analyzed to enable accurate location of harmonic peaks [5] [6]. Various linear preprocessing steps can be used to make the process of locating frequency domain features easier, such as performing linear prediction on the signal and using the residual signal for pitch detection. Performing nonlinear operations such as peak limiting also simplifies the location of harmonics. Although there are many methods of pitch estimation and tracking, both in time and frequency domains, accurate and robust detection and tracking is still a difficult problem. Most of theses methods are based on the assumption that speech signal is stationary in short time, but speech signal is non-stationary and quasi-periodical. Among these methods, autocorrelation-based method is comparatively robust against noises, but it may result in a half-pitch or double pitch error, and if noise is high, this method can't detect pitch properly. In this paper we improve the performance of robust algorithms for pitch tracking (RAPT) by means of a nonlinear preprocessing whit a filter based on mathematical morphology. The used RAPT is due to D. Talkin [7, 8] with only two minor differences.

2   The Mathematical Morphology

Mathematical morphology was proposed by J.Serra and G. Matheron in 1966, was theorized in the mid-seventies and matured from the beginning of 80's. Mathematical morphology is based on two fundamental operators: dilation and erosion. It can process binary signals and gray level signals and it has found its maximum expression in image processing applications. These two basic operations are done by means of a structuring element. The structuring element is a set in the Euclidean space and it can takes different shapes as circles, squares, or lines. Using different structuring elements it will achieve different results; therefore, the election of an appropriate structuring element is essential. A binary signal can be considered a set and dilation and erosion are Minkowski addition and subtraction with the structuring element [9]. In the context of speech processing we work with gray level signals. In this context, the addition and subtraction operations in binary morphology are replaced by suprermum and infimum operations. Moreover, on the digital signal processing framework, supremum and infimum can be changed by maximum and minimum operations.

We define the erosion as the minimum value of the part of the image function in the mobile window defined by the structuring element, Y, when its beginning is situated on x (one-dimensional framework) or in x,y (two-dimensional framework). As we deal with speech signals we are interested in one-dimensional definitions. Then, given the one-dimensional signal, the f function, and the flat structuring element, Y, the erosion can be defined as:

$$\varepsilon_Y(f)(x) = \min_{s \in Y} f(x + s) \tag{1}$$

The erosion uses the structuring element as a template, and gives the minimum gray level value of the window function defined by the mobile template; decreasing peaks and accentuating valleys (see fig.1 b).

On the other hand the gray level signals dilation is defined as:

$$\delta_Y(f)(x) = \max_{s \in Y} f(x - s) \tag{2}$$

The dilation gives the maximum gray level value of the part of the function included inside the mobile template defined by the structuring element, accentuating peaks and minimizing valleys. By combining dilation and erosion we can form other morphological operations. Opening and closing are basic morphological filters.

The morphological opening of a signal f by the structuring element Y is denoted by $\gamma_Y(f)$ and is defined as the erosion of f by Y followed of dilation by the same structuring element Y. This is:

$$\gamma_Y(f) = \delta_Y(\varepsilon_Y(f)) \tag{3}$$

And the morphological closing of a signal f by the structuring element Y is denoted by $\varphi_Y(f)$ and it is defined as the dilation of f by Y followed of the erosion by the same structuring element:

$$\varphi_Y(f) = \varepsilon_Y(\delta_Y(f)) \tag{4}$$

Opening and closing are dual operators. Closing is an extensive transform and opening is an anti-extensive transform. Both operations keep the ordering relation between two images (or functions) and are idempotent transforms. [9]. In the image context the morphological opening removes small objects from an image while preserving the shape and size of larger objects, and the morphological closing fills the gaps between objects. In the one-dimensional context both operations -by means of a non-linear process- create a more simple function than the original.

By combining an opening and a closing, both of them with the same structuring element, we can only create four different morphological filters. Then, considering the operations γ_Y and φ_Y, the four filters we could obtain are $\gamma_Y\varphi_Y$, $\varphi_Y\gamma_Y$, $\gamma_Y\varphi_Y\gamma_Y$ and $\varphi_Y\gamma_Y\varphi_Y$. From the composition of γ_Y and φ_Y no other different filter can be produced as a consequence of idempotency property.

To derive different families of morphological filters we need to combine openings and closings whit different structuring elements. There is a well known method to obtain new filters by alternating appropriately theses operators. The resulting filters are called alternating sequential filters [9] which are very effective tools to fight against noise.

In Fig.1 we can see how these morphological operators work. In Fig.1(a) we have represented a fragment of 0.1ms of speech signal, in (b) it appears the original signal (in black), a dilation (in red) and an erosion (in blue), and in (c) there are the original signal (in black), a morphological close (in red) and a morphological open (in blue). All the morphological operators involved in fig.1 use a flat structuring element of length 60 samples (3.75ms).

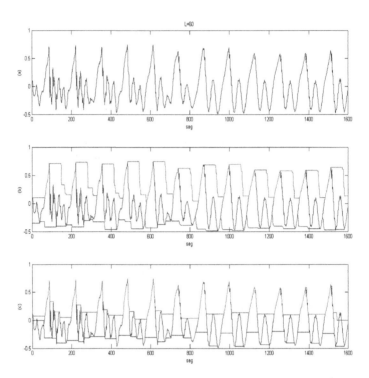

Fig. 1. (a) Original signal (b) red: dilation, blue: erosion (c) red: morphological closing, blue: morphological opening. All results obtained using a flat structuring element of L=60 (3.75ms).

In fig.2 we apply a closing with different structuring elements on the same input signal. We can appreciate the variation of the results depending on the length of the structuring element.

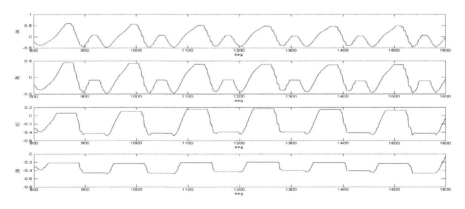

Fig. 2. Morphological closings -by structuring element length L -of the same signal (a) L=10 (b) L=20 (b) L=40 (b) L=60

3   The Selected Pitch Tracking Algorithm

In order to track the pitch we have used free software provided in the voicetoolbox for MATLAB that can be modified and redistributed under the terms of the GNU - General Public License- [8]. The Robust Algorithm for Pitch Tracking (RAPT) is taken from the work of D. Talkin [7] with only two differences. The first is related whit the modification of the coefficient AFACT which in the Talkin algorithm corresponds approximately to the absolute level of harmonic noise in the correlation window. In the used version this value is calculated as the maximum of three figures: (i) an absolute floor set, (ii) a multiple of the peak signal and (iii) a multiple of the noise floor [8]. The second difference is that the LPC used in calculating the Itakura distance uses a Hamming window rather than a Hanning window.

The software plots a graph showing lag candidates of pitch values and draws the selected path. This original signal representation could be seen in fig.4 where in the upper side there, in blue, the parts detected as voice and in red the parts detected as silent. Down, with red crosses indicating the beginning of a frame, there is represented the possible pitch values and the evolution of the selected path are depicted with continuous blue line. The pitch is given in time units (period). This pitch tracking algorithm is very robust and maintains a good performance under hard noise conditions. However, as the signal-to-noise ration increases the estimation falls into errors. To show these limitations we have introduced an additive white Gaussian noise to the same fragment of signal represented in fig. 4. In next fig. 5 we represent its behavior under three different conditions. The additive white Gaussian noise introduced in the signal has an effect on the entire voice band. In order to improve the performance of the RAPT we propose a nonlinear preprocessing filtering base on the mathematical morphology. In the left part of fig. 5 we can see the RAPT performance when the input signal has a SNR of 0,5dB the graphic shows that only some parts of the original speech are recognized as voice. In the right hand side of fig. 5 we can see the RAPT performance when the input signal has a SNR of -3.5dB; in those conditions the algorithm doesn't work.

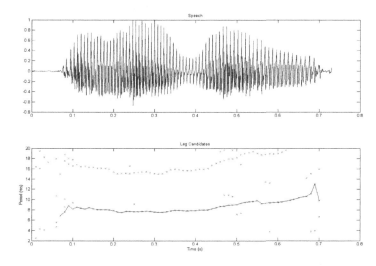

Fig. 3. Pitch evolution given by the RAPT

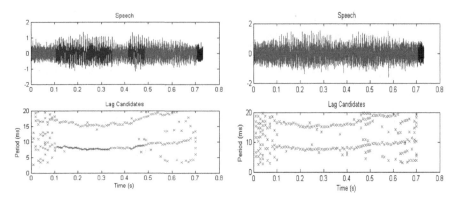

Fig. 4. Left: RAPT performance when the input signal has a SNR of 0.5 dB; only some parts of the original speech are recognized as voice. Right: RAPT performance when the input has a SNR of -3.5 dB; in those conditions the algorithm doesn't work.

4 Signal Preprocessing Based on Mathematical Morphology

In order to obtain a new representation of the noisily input signal that preserves the pitch information we propose the application of morphologic filters in a preprocessing stage. In this work we deal only with flat structuring elements. To design the appropriate filter we have explored different morphologic filters configurations with different structuring element lengths. Those studies had been done using a speech database. We have found that the input signal preprocessing by very simple filters like the compositions $\varphi_5\gamma_5$ or $\varphi_3\gamma_3$ improves the RAPT performance. Theses results could be appreciated in fig. 6 for the same fragment of the signal represented in fig. 4. In fig. 6 the signal is corrupted with Gaussian noise; in its left side we have applied this signal directly to the RAPT algorithm and in the right side we have a morphological preprocessing by $\varphi_3\gamma_3$.

More the sophisticated filters improve theses results. We propose $\varphi_4\gamma_4\varphi_3\gamma_3\varphi_2\gamma_2$.

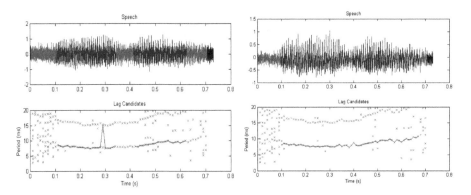

Fig. 5. RAPT performance. (Left) Input signal of SNR of 0.5 dB. (Right) The input signal of SNR of 0.5 dB had been previously preprocessed by $\varphi_3\gamma_3$.

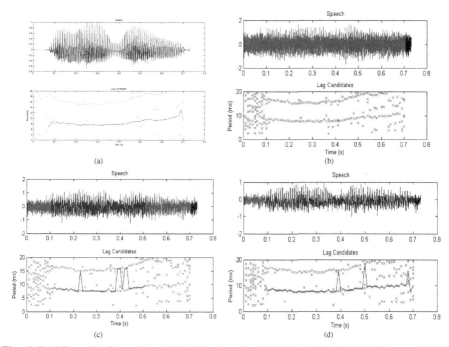

Fig. 6. RAPT results for an input: (a) without noise (b) SNR=-3.5 dB (c) SNR =   -3.5dB preprocessing by $\varphi_3\gamma_3$ (d) SNR=-3.5dB and preprocessing by $\varphi_4\gamma_4\varphi_3\gamma_3\varphi_2\gamma_2$.

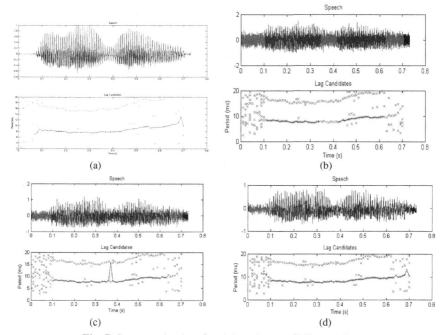

Fig. 7. Same results than fig. 6 changing the SNR to -0.5dB

5 Conclusions

In this paper we have shown how a pre-processing based on mathematical morphological filters improves the performance of pitch trackers when the input signal are corrupted with additive white Gaussian noise. As the short lengths of the structuring elements -in terms of sample time- used in the designed morphological filters it is obtained a high frequency noise reduction inside the speech band. This noise reduction has benefits in the performance of different parts of the pith tracker. One of the parts in which the improvement is important is the voice-silence detector. As this element is in the first part of the processing chain its improvement is directly reflected in the RAPT performance. The nonlinear mathematical morphology techniques are widely developed in image processing and its results can often been exported from two-dimensional to one-dimensional framework. From our knowledge those techniques are not widely explored in speech processing. In [10] we have found a work that also uses simple morphologic filters to estimate the pitch. The objective of [10] is quite different of ours: they are interested in the estimator and we are interested in the signal pre-processing. This could be reflected in the design of the morphological filters and in the size of the structuring elements that such filters use.

Acknowledgments. This work has been supported by the University of Vic under the grant R0904.

References

1. Mozziconacci, S., Hermes, D.: Study of intonation patterns in speech expressing emotion or attitude: production and perception. IPO Annual Progress Report. IPO, Eindhoven (1998)
2. Juang, B.-H., Furui, S.: Automatic recognition and understanding of spoken language—a first step towards natural human–machine communication. Proc. IEEE 88(8), 1142–1165 (2000)
3. Petrushin, V.A.: Emotion recognition in speech signal: experimental study, development, and application. In: Proc. Internat. Conf. on Spoken Language Processing, Beijing, China (2000)
4. Kang, B.-S., Han, C.H., Lee, S.-T., Youn, D.H., Lee, C.: Speaker dependent emotion recognition using speech signals. In: Proc. Internat. Conf. on Spoken Language Processing, Beijing, China (2000)
5. Hess, W.: Pitch Determination of Speech Signals. Springer, Berlin (1983)
6. Rabiner, L.R., Cheng, M.J., Rosenberg, A.E., McGonegal, C.A.: A Comparative Performance Study of Several Pitch Detection Algorithms. IEEE Trans. on Acoustics, Speech, and Signal Processing 24(5), 399–418 (1976)
7. Talkin, D.: A Robust Algorithm for Pitch Tracking (RAPT). In: Kleijn, W.B., Paliwal, K.K. (eds.) Speech Coding & Synthesis. Elsevier, Amsterdam (1995)
8. http://www.ee.ic.ac.uk/hp/staff/dmb/voicebox/voicebox.html
9. Serra, J.: Image Analysis and Mathematical Morphology. Academic, New York (1982)
10. Xiaoqun, Z., Guangyan, W.: A New Approach of the Morphology Filter for Pitch Contrail Smoothing of Chinese Tone. Signal Processing 19(4), 354–357 (2003)

Deterministic Particle Filtering and Application to Diagnosis of a Roller Bearing

Ouafae Bennis[1] and Frédéric Kratz[2]

[1] Institut PRISME – Université d'Orléans
21 rue Loigny La Bataille , 28000 Chartres, France,
[2] Institut PRISME – ENSI de Bourges
88 Boulevard Lahitolle, 18 000 Bourges, France
Ouafae.bennis@univ-orleans.fr, frederic.kratz@ensi-bourges.fr

Abstract. In this article, the detection of a fault on the inner race of a roller bearing is presented as a problem of optimal estimation of a hidden fault, via measures delivered by a vibration sensor. First, we propose a linear model for the transmission of a vibratory signal to the sensor's diaphragm. The impact of shocks due to the default is represented by a stochastic drift term whose values are in a discrete set. To determine the state of the roller bearing, we estimate the value of this term using particular filtering.

Keywords: Particular filtering, diagnosis.

1 Introduction

The diagnosis of the rotating machines became a stake major for industry, in particular aeronautical or iron and steel, because of the gigantic costs generated by a catastrophic failure and consequences on the public image of the company.

In this study, we are interested with a detection of a failure on the inner race of a roller bearing on a turbo reactor. Many sensors are given at different bearings on the reactor except on the suspected one. Hence we have a partial observation of the state of the roller from the sensors.

The physical model used is developed in [1]. The state variables we consider are respectively the displacement and the speed of the vibration sensor's diaphragm. We observe the second variable across a propagation factor α, high level noise context. Under a state representation, this is close to an optimal filtering problem.

Our aim is to estimate the hypothesis of there being a fault and subsequence level deterioration.

2 State Representation of the Problem

Let y(t) be the response of the sensor's diaphragm to an excitation h(t). The vibratory signal next to it is [1]: $y(t) = e_d(t) * h(t)$, for t>0, with $e_d(t) = \sin(\omega_r t)e^{-\frac{t}{T_c}}$, ω_r is the

J. Solé-Casals and V. Zaiats (Eds.): NOLISP 2009, LNAI 5933, pp. 171–177, 2010.

resonance pulsation, T_c is the time-constant of the sensor and * the convolution product. This obeys to the differential equation:

$$\ddot{y}(t) + \frac{2}{T_c^2}\dot{y}(t) + \left(\omega_r^2 + \frac{1}{T_c^2}\right)y(t) = \omega_r h(t)$$

Let $X(t) = [y(t) \quad \dot{y}(t)]^T$, be the displacement and the speed vector, z(t) the signal measured. For a healthy bearing, the membrane of the sensor receives the noises in the neighbourhood of the sensor, which one summarizes with a continuous Gaussian white noise w(t) with intensity Q(t). Then we have:

$$\begin{cases} \dot{X}(t) = AX(t) + Bw(t) \\ z(t) = HX(t) + v(t) \end{cases} \tag{2.1}$$

$$A = \begin{bmatrix} 0 & 1 \\ -\left(\omega_r^2 + \dfrac{1}{T_c^2}\right) & -\dfrac{2}{T_c^2} \end{bmatrix} \qquad B = \begin{bmatrix} 0 \\ \omega_r \end{bmatrix} \qquad H = [\alpha \quad 0]$$

v(t) represents the measurement noise. It is assumed that v(t) can be described as white noise where R is the variance matrix. α is a factor of propagation in the sensor's diaphragm: $0 < \alpha < 1$.

In the case of a fault on the inner race, an impulse is generated when each element of the roller comes into contact with the fault. The sensor's diaphragm receives in addition to w(t) a discontinuous component: $h_{shock}(t) = U.\sum_n \delta(t - nT_d)$

T_d and U are respectively the period of the fault and the load applied at the contact point. T_d is assumed to be deterministic and known. We choose the sampling period T_e so that: $T_d = 72.T_e$ In this case $h_{shock}(t)$ is reduced to a maximum of a single shock in the period T_e. Consequently, the fault introduces a stochastic drift term in the second state equation each time k is a multiple of 72.

U depends of the deterioration level of the bearing roller; its value grows slowly in time. Hence, we consider it stochastic with values in a discrete set $\Im = \{u_1, u_2, u_3, \ldots, u_N\}$. The filtering model is still linear Gaussian verifying:

$$\begin{cases} \dot{X}(t) = AX(t) + Bw(t) + B.U.\sum_n \delta(t - nT_d) \\ z(t) = HX(t) + v(t) \end{cases} \tag{2.2}$$

Discrete equations. The measurements z are collected at discrete times, it's more natural to deal with the discrete time system:

$$\begin{cases} X_{k+1} = FX_k + Gw_k + Gu_k^i, \\ z_k = HX_k + v_k \end{cases} \qquad k > 0 \tag{2.3}$$

$$F = (I + A.T_e), \quad G = \begin{bmatrix} 0 \\ T_e \omega_r \end{bmatrix}, \quad u_k^i = U \delta_{k,72i}$$

where δ represents the Kronecker symbol (i.e. $\delta_{a,b} = 1$ if $a = b$ $\delta_{a,b} = 0$ otherwise).

3 Roller Bearing Diagnosis

The deterioration of a roller bearing is a slow phenomenon, except in case of deficient conditions of use. We suppose that the value U is constant on a limited time interval. So a diagnosis of roller bearing comes down to estimating the possibility of shock occurrence and its amplitude U. The evolution of the amplitude U value is defined by the transition matrix P_{tr}. An element $p(i,j)$ of the matrix P_{tr} defines the probability of the change from U_i to U_j. The diagonal elements are more important than the others. The probability of a change is lower then the probability of remaining in the same state and the probability of going backward to a faultless case is close to 0. Hence P_{tr} is a triangular matrix. The diagnosis of the state of the bearing is equivalent to estimate the parameter U, the measurements z_k on the trajectory from 0 to k.

4 Particular Filtering Applied to Diagnosis

Let fix U at the value $\hat{U}_k$, least estimate of U, in (2.3). The problem is hence linear and Gaussian.

4.1 The Linear Gaussian Component Estimation

The optimal estimation of X is performed by the Kalman filter equations:

Prediction step

a priori estimate
$$\begin{cases} \hat{X}_{k|k-1} = F\hat{X}_{k-1|k-1} + G\hat{U}_k \\ P_{k|k-1} = FP_{k-1|k-1}F^T + GQ_kG^T \end{cases} \tag{4.1}$$

Correction step

a posteriori estimate
$$\begin{cases} \hat{X}_{k|k} = \hat{X}_{k|k-1} + K_k \tilde{z}_k \\ K_k = P_{k|k-1}H^T \left[HP_{k|k-1}H^T + R_k\right]^{-1} \\ P_{k|k} = [I - K_k H]P_{k|k-1} \\ \tilde{z}_k = z_k - \hat{z}_{k|k-1} \end{cases} \tag{4.2}$$

4.2 Default Detection/Estimation

To perform the diagnosis problem, we have to estimate the parameter U. This will be accomplished by the particular filtering techniques.

Every step k, we realise N deterministic draws in the set $\Im = \{u_1, u_2, u_3, \ldots, u_N\}$. The N realisations (i.e. Particles) will be used in the equation (2.3) to generate N different trajectories. For every particle U_i, we calculate the likelihood of the trajectory :

$V_k = V\left(X_k, U_k | Z_{[0;k]}\right) = \ln\left(p\left(X_k, U_k | Z_{[0;k]}\right)\right)$. The optimal estimate of U is:

$\hat{U}_k = \arg\left(\max_{U_i}(V_k)\right)$, with respect to (4.1) and (4.2).

5 Estimation/Detection Algorithm

At every step k, let explore all the U values in the set $\Im = \{u_1, u_2, u_3, \ldots, u_N\}$. The aim is to determinate the value maximising the likelihood V_k.

5.1 Initialisation

With N initial particles which are filled according to some probability measure, the initial likelihoods are: $V_0^i = \ln(p(X_0))$ for I = 1 to N.

5.2 Prediction/Correction of the Gaussian Component X

For every particle U_i, N Kalman filters are engaged to perform the estimate $\hat{X}^i_{k|k-1}$ corresponding to every U value, from the formulas (4.1), (4.2).

5.3 Evolution of the Gaussian Component X, New Particles

At the step k+1, for every initial condition $\hat{X}^i_{k|k}$, we pull M particles in the set $\Im = \{u_1, u_2, u_3, \ldots, u_N\}$, which gives M*N particles and M*N trajectories from (2.3).

5.4 Likelihood Ratio

Equations (4.1), (4.2) give a priori estimates of X and P. The innovation process leads to the calculus of V_k The system in (2.3) is markovian with independent noises, hence V_k can be derived recursively by the formulas, for i= 1 to M*N:

$$v_k^i = V_{k-1}^i + \ln\left(p\left(z_k | x_k, U_k = u_k^i\right)\right) = V_{k-1}^i + \ln\left(p\left(u_k^i\right)\right) + v_k^i$$

$$v_k^i = -\frac{1}{2}\left[\ln\left(HP_{k|k-1}H^T + R\right) + \left(z_k - H\hat{X}_{k|k-1}\right)^T \left(HP_{k|k-1}H^T + R\right)^{-1}\left(z_k - H\hat{X}_{k|k-1}\right)\right]$$

5.5 Estimation/ Detection of U

The value of U, defining the deterioration level of the roller bearing is: $\hat{U}_k = \arg\left(\max_{U_i}(V_k) \right)$, with respect to (4.1) and (4.2).

5.6 Redistribution

For the next iteration, we must sort N from M*N particles, whose have the best likelihood ratio. The algorithm will branch to **5.2.**

6 Simulations and Results

The system equations have been simulated on Matlab, with the following numerical values: $\omega_r = 7039.88$, $T_c = 0.00029$, $\alpha = 1$, $T_e = 0.000025$, $T_d = 0.0018$, $w_k \approx N(0, q_k)$, $q_k = 0.25$, $v_k \approx N(0, r_k)$, $r_k = 0.1$, for all $k \geq 0$, $X_0 \approx N(\overline{X}_0, Q_0)$, $P_0 = Q_0 = \mathrm{diag}(0,0.25)$, $\overline{X}_0 = [10 \ \ 0]^T$, X_0, w_k and v_k are assumed independents, N and M are fixed at 5 and $\mathfrak{I} = \{0.125 \quad 37 \quad 410 \quad 1010 \quad 607\}$, $P_{tr} = [995\ 3\ 2\ 0.1\ 0.1;1$ $994\ 3\ 2\ 0.1;0.1\ 1\ 995\ 3\ 1;0.1\ 1\ 1\ 995\ 3;0.1\ 0.1\ 1\ 3\ 996\ ]*1e\text{-}3$.

Case 1: the fault is assumed constant

We fix the real fault at an arbitrary value U in the interval [0.01, 10000]. The algorithm converges to the nearest value in $\mathfrak{I}$ after only 15 iterations.

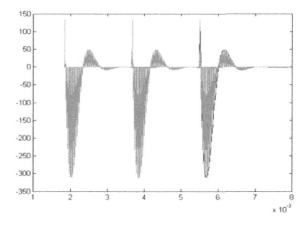

Fig. 1. Succession of shocks for a faulty roller bearing, real and estimated speed

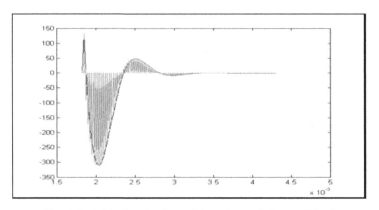

Fig. 2. Real (blue) and estimated (green) speed, Zoom of Fig 1

Case 2: The fault amplitude takes value in the set $\mathfrak{I}_{\mathbf{real}}$ = {0.1 45 450 570 950} different from $\mathfrak{I}$ ={0.125 37 410 1010 607}, the set of U particles values.

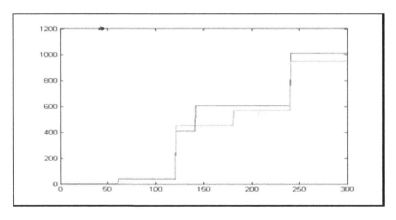

Fig. 3. Real (green) and estimated (red) default value.

7 Conclusion

In this study, the linear model proposed for the transmission of a vibratory signal to the sensor has made possible the representation of the roller bearing diagnosis by a hybrid system. The discrete parameter U representing exactly here the level of the suspected deterioration. This leaves a non linear filtering problem that we solve using particular filtering in two situations: 1°) the fault is constant, 2°) the deterioration is time varying in a discrete set of values..

References

1. McFadden, P.D., Smith, J.D.: Model for the vibration produced by a single point defect in a rolling element bearing. Journal of Sound and Vibration 96, 69–82 (1984)
2. Mohinder, S., Grewal, A.P.: Kalman Filtering, Theory and practice using Matlab. Wiley inter-science, Chichester (2001)
3. Compillo, F.: Filtrage Linéaire, Non Linéaire, et Approximation Particulaire pour le Praticien. INRIA, Mars (2005)
4. Saboni, O.: A State Representation for the Diagnosis of a Roller Bearing and Kalman Filtering. In: ICTTA 2006, DAMAS (April 2006)
5. Ben Salem, F.: Réception Particulaire Pour Canaux Multi-Trajets Evanescents en Communications Radiomobiles, Thèse de Doctorat de l'Université Paul Sabatier de Toulouse (2002)

Applications of Cumulants in Speech Processing

Vladimir Zaiats, Jordi Solé-Casals, Pere Martí-Puig, and Ramon Reig-Bolao

Digital Technologies Group, Universitat de Vic, c/. Sagrada Famlia, 7, 08500 Vic
(Barcelona), Spain
{vladimir.zaiats,jordi.sole.pere.marti,ramon.reig}@uvic.cat

Abstract. This paper analyzes applications of cumulant analysis in speech processing. The main focus is made on an integral representation for cumulants using integrals involving cyclic products of kernels. These integrals serve as a tool for statistical analysis of different statistics appearing in speech processing and speech recognition. As an example, we consider statistics based on the sample correlogram. A representation for the cumulants of these statistics using integrals involving cyclic products of kernels is obtained.

Keywords: Cumulants, higher-order statistics, correlogram, speech enhancement.

1 Introduction

Different methods in speech recognition use linear and non-linear procedures derived from the speech signal by matching the autocorrelation or the power spectrum [7,8,9]. Many of these methods perform well for clean speech, while their performance decreases strongly if noise conditions mismatch for training and testing.

As was mentioned in [9], a way for improvement of the performance of speech recognizers based on linear prediction consists in application of cumulant-based linear prediction. The recognizer obtained in [9] was shown to work well for Gaussian (white or coloured) noise.

In [4], a nonlinear parameter estimator for a noncausal autoregressive (AR) system was proposed. The clue was to derive a quadratic equation relating the unknown AR parameters to higher-order cumulants of non-Gaussian output measurements in the presence of additive Gaussian noise. This method is rather general and may be applied in speech processing.

In [10], a cumulant-based Wiener filtering (AR3_IF) was applied for robust speech recognition. A low complexity approach of this algorithm was tested in presence of bathroom water noise and its performance was compared to the classical spectral subtraction method. The method was shown to have a low sensitivity to the noise considered within the speech recognition task.

In [12], a version of the recursive instrumental variable method for adaptive parameter identification of ARMA processes was developed. In this setting, cumulant-based 'normal' equations could be obtained by means of non-conventional orthogonality conditions in the linear prediction problem. A pair of

J. Solé-Casals and V. Zaiats (Eds.): NOLISP 2009, LNAI 5933, pp. 178–183, 2010.

lattices was obtained, one excited by the observed process, and the other excited by the instrumental process. The lattice structure leaded to the AR compensated residual time series. Therefore adaptive versions of cumulant-based moving average parameter identification algorithms were directly applicable.

In [8], a new method for speech enhancement using time-domain optimum filters and fourth-order cumulants was introduced. It was based on obtaining some new properties of the fourth-order cumulants of speech signals. The analytical expression for the fourth-order cumulants was derived under the assumption that the model was sinusoidal admitting up to two harmonics per band. The main idea was to use fourth-order cumulants of the noisy speech to estimate the parameters required for the enhancement filters. The noise reduction was satisfactory in Gaussian, street, and fan noises. The effectiveness of this approach, however, was shown to diminish in harmonic and impulsive noises like office and car engine, since the discrimination between speech and noise based on fourth-order cumulants was becoming more difficult.

In [6], cumulants were used for obtaining a performance criterion in the problem of blind separation of sources. It is well-known that blind separation relies heavily on the assumption of independence of input sources. The relation of the problem of blind separation of sources to speech processing and speech recognition is straightforward.

We obtain a representation for cumulants of second-order statistics containing a special type of integrals that involve cyclic products of kernels. Our techniques are based on [1,2,3,5,11].

2   Integrals Involving Cyclic Products of Kernels

For $m \in \mathbb{N}$, define $\mathbb{N}_m := \{1, \ldots, m\}$. Assume that $(\mathbb{V}, \mathcal{F}_\mathbb{V})$ is a measurable space and $\mu_1, \ldots, \mu_m$ are σ-finite (real- or complex-valued) measures on $(\mathbb{V}, \mathcal{F}_\mathbb{V})$. For $m \in \mathbb{N}$, $m \geq 2$, consider the following integral:

$$\widehat{I}\,(K_1, \ldots, K_m; \bullet) \tag{1}$$

$$:= \int \cdots \int_{\mathbb{V}^m} \left[\prod_{p=1}^{m} K_p(v_p, v_{p+1}) \right] \bullet (v_1, \ldots, v_m) \mu_1(dv_1) \ldots \mu_m(dv_m)$$

where $v_{m+1} := v_1$. Integral (1) will be called an integral involving a cyclic product of kernels (IICPK).

We will denote:

$$\widehat{\prod_{p \in \mathbb{N}_m}} K_p(v_p, v_{p+1}) := \prod_{p=1}^{m} K_p(v_p, v_{p+1})$$

with $v_{m+1} := v_1$. This function will be called a cyclic product of kernels $K_1, \ldots, K_m$.

3   Cumulants of General Bilinear Forms of Gaussian Random Vectors

Suppose that $m \in \mathbb{N}$; $n_{j,1}, n_{j,2} \in \mathbb{N}$, $j \in \mathbb{N}_m$, and write

- $\bullet_{j,1} := (X_{j,1}(k),\ k \in \mathbb{N}_{n_{j,1}})$,   • $_{j,2} := (X_{j,2}(k),\ k \in \mathbb{N}_{n_{j,2}})$, $j \in \mathbb{N}_m$.

Assume that $\bullet_{j,1}$ and $\bullet_{j,2}$, $j \in \mathbb{N}_m$, are real-valued zero-mean random vectors and consider the following bilinear forms:

$$U_j := \sum_{k,l=1}^{n_{j,1},n_{j,2}} a_j(k,l) X_{j,1}(k) X_{j,2}(l), \quad j \in \mathbb{N}_m ,$$

where

$$\sum_{k,l=1}^{n_{j,1},n_{j,2}} := \sum_{k=1}^{n_{j,1}} \sum_{l=1}^{n_{j,2}} .$$

If we put

$$(a_j(k,l)) := \begin{pmatrix} a_j(1,1) & \cdots & a_j(1,n_{j,2}) \\ \vdots & \cdots & \vdots \\ a_j(n_{j,1},1) & \cdots & a_j(n_{j,1},n_{j,2}) \end{pmatrix}, \quad j \in \mathbb{N}_m ,$$

then for any $j \in \mathbb{N}_m$

$$U_j = \bullet_{j,1} (a_j(k,l)) \bullet_{j,1}^{\top}$$

$$= (X_{j,1}(1),\ldots,X_{j,1}(n_{j,1})) \begin{pmatrix} a_j(1,1) & \cdots & a_j(1,n_{j,2}) \\ \vdots & \cdots & \vdots \\ a_j(n_{j,1},1) & \cdots & a_j(n_{j,1},n_{j,2}) \end{pmatrix} \begin{pmatrix} X_{j,2}(1) \\ \vdots \\ X_{j,2}(n_{j,2}) \end{pmatrix} .$$

Consider the joint simple cumulant cum $(U_1,\ldots,U_m)$ of the random variables $U_1,\ldots,U_m$ assuming that this cumulant exists. By general properties of the cumulants, we obtain

$$\text{cum } (U_1,\ldots,U_m)$$

$$= \sum_{k_{1,1},k_{1,2}=1}^{n_{1,1},n_{1,2}} \cdots \sum_{k_{m,1},k_{m,2}=1}^{n_{m,1},n_{m,2}} \left[\left(\prod_{j=1}^{m} a_j(k_{j,1},k_{j,2}) \right) \right.$$

$$\left. \times \text{ cum } (X_{j,1}(k_{j,1})X_{j,2}(k_{j,2}), j \in \mathbb{N}_m) \right] .$$

Since any general bilinear form can be represented as a sum of diagonal bilinear forms, the following result holds.

Theorem 1. *Let* $m \in \mathbb{N}$; $n_{j,1} = n_{j,2} = n_j \in \mathbb{N}$, $j \in \mathbb{N}_m$. *Assume that* $(\bullet_{j,1},$ $\bullet_{j,2}, j \in \mathbb{N}_m)$ *is a jointly Gaussian family of zero-mean random variables and suppose that for any* $j, \tilde{j} \in \mathbb{N}_m$ *and any* $\alpha, \tilde{\alpha} \in \{1, 2\}$ *there exists a complex-valued measure* $M_{j,\tilde{j}}^{\alpha,\tilde{\alpha}}$ *such that*

$$\mathsf{E} X_{j,\alpha}(k) X_{\tilde{j},\tilde{\alpha}}(\tilde{k}) = \int_{\mathbb{D}} e^{i(k-\tilde{k})\lambda} M_{j,\tilde{j}}^{\alpha,\tilde{\alpha}}(d\lambda) \ .$$

Then

$$\mathrm{cum}\,(U_1, \ldots, U_m)$$

$$= \sum_{l \in \mathcal{L}_m(n_j, j \in \mathbb{N}_m)} \sum_{(\boldsymbol{j},\boldsymbol{\alpha}) \in \{P,2\}_{m-1}} \int_{\mathbb{D}^m} \cdots \int_{\mathbb{D}^m} \left[\widehat{\prod_{p \in \mathbb{N}_m}} K_p^{(\boldsymbol{j},\boldsymbol{\alpha},l)}(v_p - v_{p+1}) \right]$$

$$\times \mu_1^{(\boldsymbol{j},\boldsymbol{\alpha},l)}(dv_1) \ldots \mu_m^{(\boldsymbol{j},\boldsymbol{\alpha},l)}(dv_m) \ ,$$

that is $\mathrm{cum}\,(U_1, \ldots, U_m)$ *is represented as a finite sum of integrals involving cyclic products of kernels. Here,* $\bullet := (j_1, j_2, \ldots, j_m)$, $\bullet := (\alpha_1, \alpha_2, \ldots, \alpha_m)$, $j_{m+1} = j_1 = 1$, $\alpha_{m+1} = \alpha_1 = 2$, *and the sum* $\sum_{(\boldsymbol{j},\boldsymbol{\alpha})}$ *is extended to all*

$$((j_2, \ldots, j_m), (\alpha_2, \ldots, \alpha_m)) \in \mathrm{Perm}\{2, \ldots, m\} \times \{1, 2\}^{m-1} \ . \tag{2}$$

The notation $(\bullet, \bullet) \in \{P, 2\}_{m-1}$ *for fixed* $j_1 = 1$ *and* $\alpha_1 = 2$ *is equivalent to (2).*

Here, we put $\mathbb{Z}_{|n_j - 1|} := \{-(n_j - 1), \ldots, -1, 0, 1, \ldots, n_j - 1\}$ for $j \in \mathbb{N}_m$ and $\mathcal{L}_m(n_j, j \in \mathbb{N}_m) := \mathbb{Z}_{|n_1 - 1|} \times \cdots \times \mathbb{Z}_{|n_m - 1|}$.

4   Applications

We apply the above obtained integral representations to some problems in speech recognition. Let us consider a setting where sample correlograms and sample cross-correlograms of stationary time series appear.

Let $\bullet(t) := (Y_1(t), Y_2(t))$, $t \in \mathbb{Z}$, be a weak sense stationary zero-mean bidimensional vector-valued stochastic process with real-valued components whose matrix-valued autocovariance function is as follows:

$$\mathbf{C}_{\mathbf{Y}}(t) := \begin{pmatrix} C_{11}(t) & C_{12}(t) \\ C_{21}(t) & C_{22}(t) \end{pmatrix}, \quad t \in \mathbb{Z} \ ,$$

and let

$$\mathbf{F}_{\mathbf{Y}}(\lambda) := \begin{pmatrix} F_{11}(\lambda) & F_{12}(\lambda) \\ F_{21}(\lambda) & F_{22}(\lambda) \end{pmatrix}, \quad \lambda \in [-\pi, \pi] \ ,$$

be the matrix-valued spectral function of the vector-valued process $\bullet\ (t), t \in \mathbb{Z}$. Let $\gamma, \delta \in \{1, 2\}$. Consider the following random variables:

$$\hat{C}_{\gamma\delta}(\tau; N) := \sum_{k=1}^{N} b_{\gamma\delta}(k;\ \tau, N)Y_\gamma(k+\tau)Y_\delta(k), \quad \tau \in \mathbb{Z}, \quad N \in \mathbb{N},$$

where $b_{\gamma\delta}(k;\ \tau, N)$, $k \in \mathbb{N}_N$, $\tau \in \mathbb{Z}$, $N \in \mathbb{N}$, are non random real-valued weights. It is often assumed that

$$\sum_{k=1}^{N} b_{\gamma\delta}(k;\ \tau, N) = 1, \quad \tau \in \mathbb{Z}, \quad N \in \mathbb{N}. \tag{3}$$

For example, let $N \in \mathbb{N}$ be given and let

$$b_{\gamma\delta}(k;\ \tau, N) = \frac{1}{N}, \quad k \in \mathbb{N}_N, \quad \tau \in \mathbb{Z}.$$

Then (3) holds and

$$\hat{C}_{\gamma\delta}(\tau; N) = \frac{1}{N} \sum_{k=1}^{N} Y_\gamma(k+\tau)Y_\delta(k), \quad \tau \in \mathbb{Z}, \quad N \in \mathbb{N}.$$

The following sample correlograms are also often used in spectral analysis and speech recognition as estimates of $C_{\gamma\delta}(\cdot)$, $\gamma, \delta \in \{1, 2\}$:

$$\tilde{C}_{\gamma\delta}(\tau; N) = \begin{cases} \dfrac{1}{N} \displaystyle\sum_{k=1}^{N-|\tau|} Y_\gamma(k+|\tau|)Y_\delta(k), & \text{for } |\tau| < N; \\ 0, & \text{for } |\tau| \geq N, \end{cases}$$

$$\tilde{\tilde{C}}_{\gamma\delta}(\tau; N) = \begin{cases} \dfrac{1}{N-|\tau|} \displaystyle\sum_{k=1}^{N-|\tau|} Y_\gamma(k+|\tau|)Y_\delta(k), & \text{for } |\tau| < N; \\ 0, & \text{for } |\tau| \geq N. \end{cases}$$

Let $\gamma, \delta \in \{1, 2\}$, $N \in \mathbb{N}$, $m \in \mathbb{N}$, and $\tau_j \in \mathbb{Z}$, $j \in \mathbb{N}_m$. Put

$$\mathrm{cum}_{\gamma\delta}^{(N)}(\tau_1, \ldots, \tau_m) := \mathrm{cum}(\hat{C}_{\gamma\delta}(\tau_j; N),\ j \in \mathbb{N}_m);$$

$$n_j = N, \quad j \in \mathbb{N}_m; \qquad a_j(k) = b_{\gamma\delta}(k;\ \tau_j, N), \quad k \in \mathbb{N}_N,\ j \in \mathbb{N}_m;$$

$$X_{j,1}(k) = Y_\gamma(k+\tau_j), \quad k \in \mathbb{N}_N,\ j \in \mathbb{N}_m;$$

$$X_{j,2}(k) = Y_\delta(k), \quad k \in \mathbb{N}_N,\ j \in \mathbb{N}_m.$$

Under these conditions the results obtained in Section 3 can be applied to the cumulants. These results imply that the Gaussian component of the cumulant $\mathrm{cum}_{\gamma\delta}^{(N)}(\tau_1, \ldots, \tau_m)$ is represented as a finite sum of integrals involving cyclic products of kernels.

Acknowledgements. We acknowledge the support of the Universitat de Vic under grants R0904 and R0912. Some parts of this paper are based on joint research with V. Buldygin (National Technical University of Ukraine) and F. Utzet (Universitat Autnoma de Barcelona).

References

1. Avram, F.: Generalized Szego Theorems and Asymptotics of Cumulants by Graphical Methods. Trans. Amer. Math. Soc. 330, 637–649 (1992)
2. Avram, F., Taqqu, M.: On the Generalized Brascamp-Lieb-Barthe Inequality, a Szegő Type Limit Theorem, and the Asymptotic Theory of Random Sums, Integrals and Quadratic Forms (2005) (manuscript),
   http://www.univ-pau.fr/~avram/papers/avramtaqqu.pdf
3. Buldygin, V., Utzet, F., Zaiats, V.: Asymptotic Normality of Cross-Correlogram Estimators of the Response Function. Statist. Inf. Stoch. Process. 7, 1–34 (2004)
4. Chi, C.-Y., Hwang, J.-L., Rau, C.-F.: A New Cumulant Based Parameter Estimation Method for Noncausal Autoregressive Systems. IEEE Trans. Signal Process. 42, 2524–2527 (1994)
5. Grenander, U., Szegő, G.: Toeplitz Forms and Their Applications. University of California Press, San Francisco (1958)
6. Ihm, B.-C., Park, D.-J.: Blind Separation of Sources Using Higher-Order Cumulants. Signal Process. 73, 267–276 (1999)
7. Makhoul, J.: Linear Prediction: A Tutorial Review. Proc. IEEE 63, 561–580 (1975)
8. Nemer, E., Goubran, R., Mahmoud, S.: Speech Enhancement Using Fourth-Order Cumulants and Optimum Filters in the Subband Domain. Speech Comm. 36, 219–246 (2002)
9. Paliwal, K.K., Sondhi, M.M.: Recognition of Noisy Speech Using Cumulant-Based Linear Prediction Analysis. In: International Conference on Acoustics, Speech and Signal Processing, ICASSP 1991, April 14–17, pp. 429–432 (1991)
10. Salavedra, J.M., Hernando, H.: Third-Order Cumulant-Based Wiener Filtering Algorithm Applied to Robust Speech Recognition. In: Ramponi, G., Sicuranza, G.L., Carrato, S., Marsi, S. (eds.) 8th European Signal Processing Conference, EUSIPCO 1996, Trieste, Italy, September 10-13 (1996)
11. Sugrailis, D.: On Multiple Poisson Stochastic Integrals and Associated Markov Semigroups. Probab. Math. Statist. 3, 217–239 (1984)
12. Swami, A., Mendel, J.M.: Adaptive System Identification Using Cumulants. In: International Conference on Acoustics, Speech and Signal Processing, ICASSP 1988, New York, NY, USA, vol. 4, pp. 2248–2251 (1988)

The Growing Hierarchical Recurrent Self Organizing Map for Phoneme Recognition

Chiraz Jlassi, Najet Arous, and Noureddine Ellouze

Ecole Nationale d'Ingénieurs de Tunis
BP 37, Belvédère 1002 Tunis, Tunisie
Chiraz_jlassi@yahoo.fr,
Najet.Arous@enit.rnu.tn

Abstract. This paper presents a new variant of a well known competitive learning algorithm: Growing Hierarchical Recurrent Self Organizing Map (GH_RSOM). The proposed variant is like the basic Growing Hierarchical Self Organizing Map (GHSOM), however, in the GH_RSOM each map of each layer is a recurrent SOM (RSOM) it is characterized for each unit of the map by a difference vector which is used for selecting the best matching unit and also for adaptation of weights of the map. In this paper, we study the learning quality of the proposed GHSOM variant and we show that it is able to reach good vowels recognition rates.

Keywords: Artificial neural network, growing hierarchical self-organizing map, recurrent self-organizing map, phoneme recognition.

1 Introduction

Spontaneous speech production is a continuous and dynamic process. This continuity is reflected in the acoustics of speech sounds and, in particular, in the transitions from one speech sound to another [1].

For a neural network to be dynamic it must be given a memory. A traditional way of building short-term memory into the structure of a neural network that has no capacity for representing temporal dependencies is through the use of time delays, which can be implemented at the input layer of the network [2].

While RSOM [3] is a promising structure in temporal sequence processing area, it is difficult to decide an appropriate network structure for a given problem. Since a fixed network structure is used in terms of number and arrangement of neurons, which has to be defined prior to training, this often leads to a significant degree of trial and error when deploying the model. Therefore, previously proposed growing neural network methods [4], [5], [6] motivate the idea of a Growing hierarchical Recurrent Self-Organizing Map (GH_RSOM). The contribution of this work is to design a hierarchical model that composed of independent RSOMs (many RSOM), each of which is allowed to grow in size during the unsupervised training process until a quality criterion regarding data representation is met.

The basic Growing Hierarchical Self Organizing Map (GHSOM) [7] use a hierarchical structure of multiple layers where each layer consists of a number of

J. Solé-Casals and V. Zaiats (Eds.): NOLISP 2009, LNAI 5933, pp. 184–190, 2010.

independent SOMs (many SOM). Only one SOM is used at the first layer of the hierarchy. For every unit in this map a SOM might be added to the next layer. This principle is repeated with the third and any further layers of the GHSOM.

The GHSOM grows in two dimensions: in width (by increasing the size of each SOM) and in depth (by increasing the number of layers) [8],[9].

In this paper, we are interested in phoneme classification by means of a growing hierarchical recurrent self-organizing map. We have used the DARPA TIMIT speech corpus for the evaluation of the GH_RSOM in domain application. We demonstrate that the proposed GHSOM variant provide more accurate phoneme classification rates in comparison with the basic GHSOM model.

In the following, we present the basic model of the GHSOM. Thereafter, we present the proposed competitive learning algorithm. Finally, we illustrate experimental results in classification of vowels of TIMIT database.

2 Basic GHSOM Model

The GHSOM enhances the capabilities of the basic SOM in two ways. The first is to use an incrementally growing version of the SOM, which does not require the user to directly specify the size of the map beforehand; the second enhancement is the ability to adapt to hierarchical structures in the data. Prior to the training process a "map" in layer 0 consisting of only one unit is created. This unit's weight vector is initialized as the mean of all input vectors and its mean quantization error (MQE) is computed. The MQE of unit i is computed as:

$$MQE_i = \frac{1}{|U_i|} \sum_{k \in U_i} \|x_k - m_i\| \qquad U_i = \{k / c_k = i\} \qquad (1)$$

Beneath the layer 0 map a new SOM is created with a size of initially 2×2 units. The intention is to increase the map size until all data items are represented well. A mean of all MQE_i is obtained as <MQE>. The <MQE> is then compared to the MQE in the layer above, <MQE>above. If the following, inequality is fulfilled a new row or column of map units are inserted in the SOM,

$$MQE > \tau_1 \cdot \langle MQE \rangle_{above} \qquad (2)$$

Where τ_1 is a user defined parameter. Once the decision is made to insert new units the remaining question is where to do so. In the GHSOM array, the unit i with the largest MQE_i is defined as the error unit. Then the most dissimilar adjacent neighbor, i.e., the unit with the largest distance in respect to the model vector, is selected and a new row or column is inserted between these. If the inequality (3) is not satisfied, the next decision to be made is if some units should be expanded on the next hierarchical level or not. If the data mapped onto one single unit i still has a larger variation, i.e.

$$MQE_i > \tau_2 \langle MQE \rangle_{above} \qquad (3)$$

Where τ_2 is a user defined parameter, then a new map will be added at a subsequent layer.

Generally, the values for τ_1 and τ_2 are chosen such that $1 > \tau_1 \gg \tau_2 > 0$. In [10] the GHSOM parameter, τ_1 and τ_2 are called "breadth"- and "depth"-controlling parameters, respectively. Generally, the smaller the parameter τ_1, the larger the SOM arrays will be. The smaller the parameter τ_2, the more layers the GHSOM will have in the hierarchy.

3 The Proposed Competitive Learning Algorithm

Speech production is a continuous and dynamic process. So the proposed growing hierarchical recurrent self organizing map GH_RSOM use a hierarchical structure of multiple layers where each layer consists of a number of independent maps. Each unit of the map defines a difference vector which is used for selecting the best matching unit and also for adaptation of weights of the map. Weight update is similar to the SOM algorithm, except that weight vector vectors are moved towards recursive linear sum of past difference vectors and the current input vector.

In the GH_RSOM each map of each layer is a recurrent SOM (RSOM). In the training algorithm an episode of consecutive input vectors x(n) starting from a random point in the input space is presented to the first map of the hierarchy. The difference vector yi(n) in each unit of the map is updated as follows:

$$y_i(n) = (1 - \alpha)y_i(n-1) + \alpha(x(n) - w_i(n)) \qquad (4)$$

where $w_i(n)$ is the weight vector, $x(n)$ is the input pattern, $0 < \alpha \leq 1$ is the memory coefficient and $y_i(n)$ is the leaked difference vector for the unit i at step n while large α corresponds to short term memory whereas small values describe long term memory, or a slow activation decay. Each unit involves an exponentially weighted linear recurrent filter with the impulse response.

$h(K) = \alpha (1 - \alpha)^k$, $k \geq 0$ see Fig.1. At the end of the episode (step n), the best matching unit (bmu) b is searched by

$$y_b = \min_{i \in V} \|y_i(n)\| \qquad (5)$$

where V is the set of all neurons comprising the RSOM. Then, the map is adopted with a modified Hebbian training rule as

$$w_i(n+1) = w_i(n) + \gamma(n)h_{i,b}(n)y_i(n,\alpha) \qquad (6)$$

where $h_{i,b}(n)$ is the value of the neighborhood function and $0 < \gamma(n) \leq 1$ is the monotonically decreasing learning rate factor.

This process is repeated layer by layer using knowledge about the BMU of the frozen layer (l-1) in the search of the BMU on the next layer (l). For example the search of the BMU in the second layer is restricted into the map connected to BMU of the first layer. And when we are in the last level of the hierarchy, we look for the label of the last BMU.

Fig 2 show a GH_RSOM hierarchy in which the first layer has 3 units expanded in the next layer of the hierarchy.

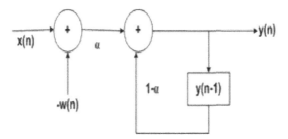

Fig. 1. Recurrent self-organizing map (RSOM)

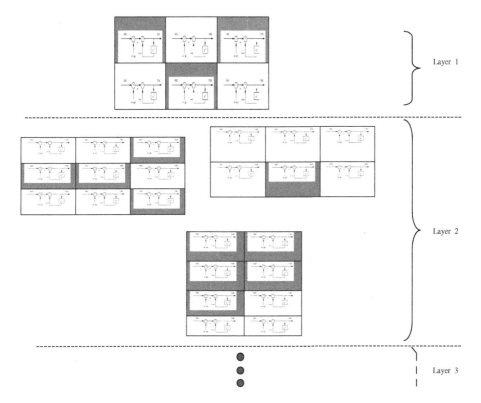

Fig. 2.   Growing hierarchical recurrent self organizing map (GH_RSOM)

4   Experimental Results

We have implemented the GHSOM variant named GH_RSOM and made experiments on voiced segment of continuous speech. The system is composed of three main components: first, signal filtering, vowels mel cepstrum coefficients production. The input space is composed by 16 ms frames of 12 mel cepstrum coefficients. 3 middle frames are selected for each phoneme. The second component is the GH_RSOM learning module. The third component is the phoneme classification module.

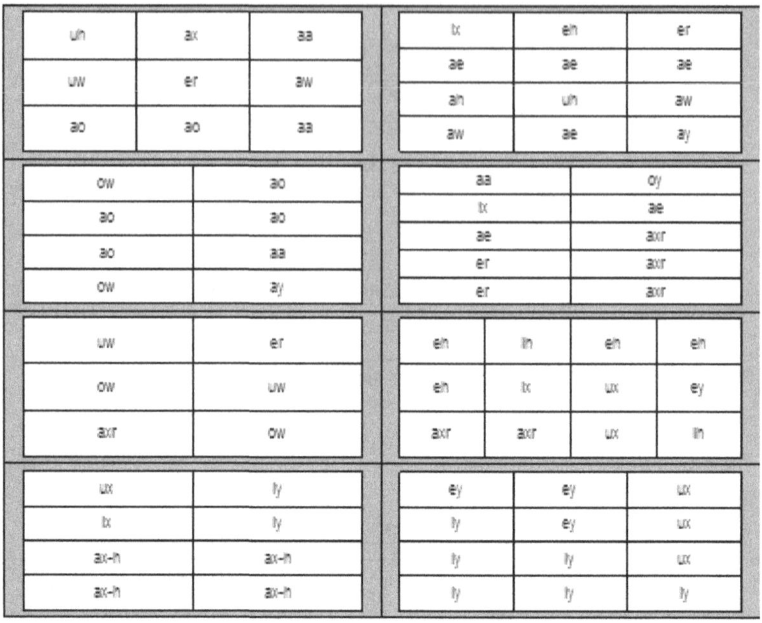

uh	ax	aa
uw	er	aw
ao	ao	aa

ix	eh	er
ae	ae	ae
ah	uh	aw
aw	ae	ay

ow		ao
ao		ao
ao		aa
ow		ay

aa		oy
ix		ae
ae		axr
er		axr
er		axr

uw		er
ow		uw
axr		ow

eh	ih	eh	eh
eh	ix	ux	ey
axr	axr	ux	ih

ux		iy
ix		iy
ax-h		ax-h
ax-h		ax-h

ey	ey	ux
iy	ey	ux
iy	iy	ux
iy	iy	iy

Fig. 3. GHSOM with 2 layers and 9 Maps labelled by the vowels of TIMIT database

iy	ix	iy
eh	iy	iy
eh	ix	iy

iy		ix
ix		ux
iy		iy
ux		ix
ux		ix

ey	iy	ux
ix	ix	iy
iy	iy	iy

ey	ih	eh
ey	iy	ix
iy	eh	ah
iy	ix	ey
eh	ix	ix
ix	ix	iy

ix	ax	er	axr	axr	ax
ae	ix	ix	eh	ux	ix
ax	ix	ix	ix	ix	ux

ow	oy	er
ax	uw	axr
ao	ah	ah
ao	aa	ae

axr	ix	ay	ae
ix	ay	aa	ay
ah	eh	aw	ae
eh		eh	ah
ah		ah	eh
ae	ay	ae	ae
ae	eh	ay	ae
eh	ow	ay	ay
aa		ay	ae

oy	ax	axr
ao	axr	axr
ax	ow	eh
ax	ix	axr

aa	aa	ah
ao	aa	ay
ao	ay	ah
ow	ao	ao

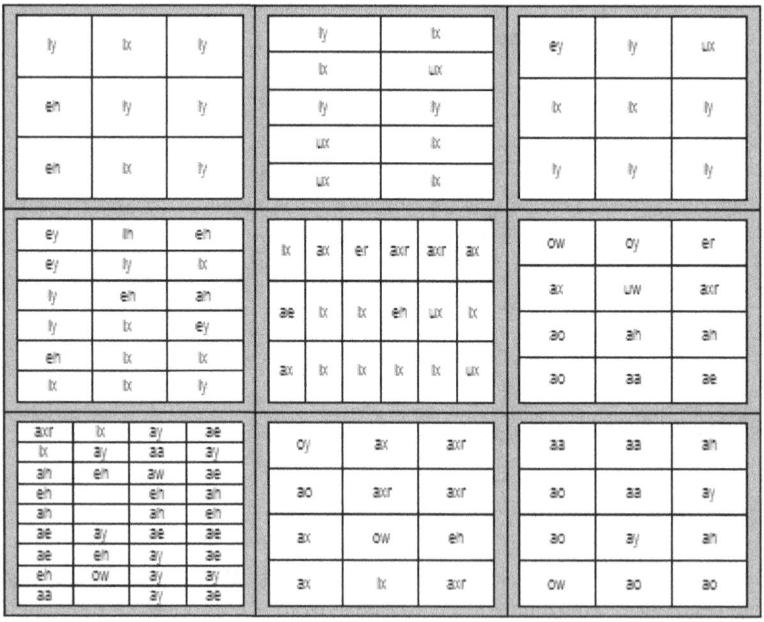

Fig. 4. GH_RSOM with 2 layers and 10 Maps labelled by the vowels of TIMIT database

Table 1. Classification rates

Phoneme	GHSOM	GH_RSOM
aw	21,45	25,92
oy	57,14	47,61
uh	19,13	61,11
uw	45,83	41,66
axh	43,11	50;98
iy	77,04	83,60
ih	14,00	25,45
eh	56,58	62,01
ey	35,00	35,33
ae	64,51	59,13
aa	16,66	51,19
ay	23,18	52,17
ah	38,75	45,00
ao	69,44	71,29
ow	15,87	42,85
ux	21,66	33,33
er	33,33	50,98
ax	37,36	32,69
ix	58,21	59,50
axr	50,00	70,19
Mean classification rates	**39,91**	**50,09**

TIMIT corpus was used to evaluate the proposed GH_RSOM in continuous speech and speaker independent context. TIMIT database contains a total of 6300 sentences, 10 sentences spoken by each of 630 speakers from 8 major dialect regions. The data is recorded at a sample rate of 16 KHz at 16 bits per sample.

In experiments, we used the New England dialect region (DR1) composed of 31 males and 18 females. This corpus contains 14399 phonetic units for training. Each phonetic unit is represented by three frames selected at the middle of each phoneme to generate data vectors.

Two experiments were conducted for different model. In the first, learning algorithm is a basic GHSOM, and we generated a 2-layers model with 8 maps in the second layer (fig 3) in the second, learning algorithm is the proposed GH_RSOM in this case a model of 2-layers with 9 maps in the second layer was generated (fig 4). In the two experiments parameter r_1 which controls the actual growth process of the GH_RSOM is equal to 0.8 and r_2 parameter controlling the minimum granularity of data representation is equal to 0,02. All maps are trained for 500 iterations and the memory coefficient is equal to 0,35 . Table 1 presents vowel classification rates for the two experiments. And good vowel recognition rates accuracy is obtained. We should note that good results are obtained even for different input frames in training phase, this prove robustness of the GH_RSOM.

5 Conclusion

In this paper, we have proposed new variant of the hierarchical neural network algorithm in the unsupervised learning category, and we are interested in phoneme recognition by means of a GHSOM variant where each neuron of each map level is characterized by a difference vector which is used for selecting the best matching unit and also for adaptation of weights of the map.

The proposed GH_RSOM provides best classification rates in comparison with the basic GHSOM model.

As a future work, we suggest to study different learning variants of the recurrent GHSOM in domain application.

References

1. Santiage, F., Alex, G., Jurgen, S.: Phoneme Recognition in TIMIT with BLSTM-CTC (2008)
2. Lichodzijewski, P., Zincir-Heywood, A.N., Heywood, M.I.: Host-Based Intrusion Detection Using Self-Organizing Maps. In: IEEE International Joint Conference on Neural Networks, pp. 1714–1719 (2002)
3. Varsta, M.J., Heikkonnen, J.D., Millan, R.: Context Learning with the Self-Organizing Map. In: Proceedings of Workshop on Self-Organizing Maps, Helsinki University of Technology, pp. 197–202 (1997)
4. Dittenbach, M., Merkl, D., Rauber, A.: The Growing Hierarchical Self- Organizing Map, Neural Networks, 2000. In: IJCNN 2000, Proceedings of the IEEE-INNS-ENNS International Joint Conference on Neural Networks, vol. 6, pp. 15–19 (2000)
5. Zhou, J., Fu, Y.: Clustering High-Dimensional Data Using Growing SOM. In: Wang, J., Liao, X.-F., Yi, Z. (eds.) ISNN 2005. LNCS, vol. 3497, pp. 63–68. Springer, Heidelberg (2005)
6. Liu, Y., Tian, D., Li, B.: A Wireless Intrusion Detection Method Based on Dynamic Growing Neural Network. In: First International Multi-Symposiums on Computer and Computational Sciences (IMSCCS 2006), vol. 2, pp. 611–615 (2006)
7. Dittenbach, M., Merkl, D., Rauber, A.: The growing hierarchical self-organizing map. In: Proceedings of the International Joint Conference on Neural Networks (IJCNN), vol. 6, pp. 15–19 (2000)
8. Dittenbach, M., Rauber, A.: Polzlbauer: Investigation of alternative strategies and quality measures for controlling the growth process of the growing hierarchical self-organizing map. In: Proceeding of the International Joint Conference on Neural Networks (IJCNN 2005), pp. 2954–2959 (2005)
9. Tangsripairoj, S., Samadzadeh, M.H.: Organizing and visualizing software repositories using the growing hierarchical self- organizing map. Journal of Information Science and Engineering 22, 283–295 (2006)
10. Pampalk, E., Widmer, G., Chan, A.: A new approach to hierarchical clustering and structuring of data With self-organizing maps. Intelligent Data Analysis Journal 8(2), 131–149 (2004)

Phoneme Recognition Using Sparse Random Projections and Ensemble Classifiers

Ioannis Atsonios

Johns Hopkins University
yannis@cs.jhu.edu

Abstract. Speech recognition is among the harderst engineering problems,it has drawn the attention of various researchers over a wide range of fields. In our work we deviate from the mainstream methods by proposing a mixture of feature extraction and dimensionality reduction method based on Random Projections that is followed by widely used non-linear and probabilistic learning method,Random Forests that has been used successfully for high dimensional data in various applications of Machine Learning. The methodological strategy decouples the problem of speech recognition to 3 distinct components: *a)feature extraction, b)dimensionality reduction,c)classification scheme*,since tackles the problem via Statistical Learning Theory perspective enriched by the current advances of Signal Processing.

Keywords: Phoneme Recognition,Random Forests,Random Projections.

1   Introduction

Speech recognition is traditionally among the hardest challenges of engineering tasks that researchers face and even if is historically open for almost 50 years,unfurtunatelly the proposed solutions in the literature are far from the superior performancy of natural "systems" that humans and animals exhibit.Albeit great progress has emerged the last years and was achieved not only to the retrofitness of intelligence but was enriched with the better understanding of diverse fields of science and engineering,namely Statistics,Signal Processing,Non-linear mathematics,Optimization,Cognitive Science,etc that all of them overlap for unlocking the latent complexity of speech recognition.However modern trends for attacking the problem are based on more unusual methodologies,the reason is not because of personal quirk of researchers,but primarily because of the advent of more powerful computers ,but more interestingly due to the craft of more sophisticated algorithmic tools that propeled the understanding of the problem.Of course this created a lively interface among scientists of various fields,and central role in this interaction was enhanced via the efforts of Cognitive scientists,since they provided better insight of how speech processing works in 'natural systems',namely the speech production mechanism,auditory processing and of course they scratched the surface of how brain is able to couple all these highly heterogeneous components.

J. Solé-Casals and V. Zaiats (Eds.): NOLISP 2009, LNAI 5933, pp. 191–198, 2010.

2   Related Work

Of great importance in any automatic speech recognition system is the feature extraction component that is followed by the learning algorithm.Every of the aforementioned modules of a speech recognition system carries a specific task,in particular feature extraction intuitively can be cast as dimensionality reduction over the raw speech data,which of course cannot be used without any preprocessing to feed the learning algorithm,primarily due to the high initial dimensionality which makes the learning task more difficult,hence is necessary to represent a faithfully in terms of latent information a speech frame via vectors with reduced dimensionality.For this reason ,mainstream methodology dictates the usage of MFCC,LPC,RASTA-PLP coefficients extended even with dynamic features(practically the derivatives of the those features in order to encapsulate the velocity and acceleration of speech signal,especially in MFCC case),this extension of using more features comes with the expense of increasing the dimensionality.Now after the feature extraction step then machine learning algorithm is needed for the classification task. Generally speech recognition has been tackled via a wealth of algorithms that is dictated by different philosophy.Namely it has been used in literature Neural Networks[1],Support Vector Machines[2],Hidden Markov Models[3],etc.All of the methods that have been proposed have various degrees of success,that is dependent over the features that have been used ,the datasets for evaluation and of course something that is among the hardest things,which is dependent on the parameters of tweaking the learning algorithms and sometimes can be extremely painful and is a topic of research on its own that has tantamount importance theoretically and practically.Neural networks are very hard to tweak and one fundamental calamity has to do with the lack of rigorous mathematical foundations,on the other hand Support Vector Machines have solid mathematical background on Optimization theory ,but practically are computationally expensive.Hidden Markov Models are the flagship of commercial systems and have been studied thoroughly in academia,especially due to their probabilistic interpretation and the existence of rigorous algorithms for training them ,namely Expectation-Maximation algorithm.

3   Proposed Technique

Main gist of our methodology is to apply non-linear feature extraction for speech data with a supervised machine learning scheme.Non-linear speech features is a great choice since potentially they can be more robust to the presence of noise,especially of noise that obeys even to non-Gaussian distributions. More specifically as feature extraction we would like to explore the coupling of mainstream methods(namely MFCC[4] and RASTA-PLP[5,6] with dynamic features) with non-linear methods,in particular as candidate for non-linear features we would like to explore the usage of Lyapunov eigenvalues[7,8],fractal dimension and higher order statistics(especially 3rd and 4th and 5th moment of signal)[9].The reason of doing this step is essentially to combine the merits of linear and non-linear methods,especially in case of noise ,non-linear methods tend to be a good choice,also has the distinctive characteristic of combining time-domain features and frequency domain features(frequency domain tend to be more robust and less sensitive to the presence of noise and is easier to manipulate signal).However even if this initial idea seems quite appealing,the computational cost can be quite prohibitive for real time applications,hence we can do an additional dimensionality reduction over the features or to feed directly the learning algorithm.For the dimensionality reduction we propose the usage of the following method:

– we can choose the idea of random projections,which is computationally very efficient and they have very strong probabilistic foundations and guarantees,in particular due to Johnson-Lindestrauss lemma,if we have n points in k-dimensional space then we can project to a space O(logn) and keep the pairwise distances within a small epsilon(small variance) with high probability.This method has been practically been a very good surrogate for PCA method,since for high dimensional data PCA can be computationally very expensive.On the other hand random projections can be very efficiently computed[10].

In the following sections we provide the decoupled structure of our algorithmic proposal:

3.1   Feature Extraction

In our work we experimented with the well studied and widely used MFCC coefficients[2].In parallel as features we computed the Δ and $\Delta\Delta$ coefficients of MFCC in order to encapsulated the velocity and acceleration of speech.MFCC is relatively easy to compute(even in real time frame).Uses Mel frequency scale which is perceptually motivated.It has strong modelling characteristics that try to decouple source and filter. Of course there are various alternatives that can be used,to name few Perceptual Linear Predictive(PLP),RASTA,etc

Computation of Δ and $\Delta\Delta$ features : Derivatives of these coefficients can help to capture the temporal information of speech.Intuitevely is like computing the time derivative of a high dimensional function per variable,more specifically we have : $\Delta X_t = X_t - X_{t-1}$ and $\Delta\Delta X_t = \Delta X_t - \Delta X_{t-1}$,at frame t.However this computation is very rough and is used primarily because of the very easy computation and because it can be done in an on-line fashion(on the fly,without having to store all the signal).Is rough in the sense since is behaving like a high frequency filter,hence if a signal x_t^i is noisy this computation makes it even more noisy.In these situations is much better the usage of *splines*,that have the ability to compute in a more global way the time derivative,albeit this computation is harder and also it has a lot of parameters(degrees of freedom) that should be tuned wisely.It is relatively easy to see that the computation of derivatives has time complexity $O(n)$.

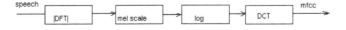

Fig. 1. MFCC computation diagram

3.2 Random Projections

Fundamental element of the majority of the methods that use random projections method in the way that are crafted is the following lemma which is widely known as *Johnson-Lindenstrauss lemma*.

Lemma 1. *For any $0 \le \epsilon \le 1$ and any integer n, let k be a positive integer that satisfies $k \ge 4ln(n)(\epsilon^2/2 - \epsilon^3/3)^{-1}$. Then for any n points in R^d, there is a map from R^d to R^k such that all the distances between mapped points f(u) and f(v) (u and v $\in R^d$), satisfy the following relation:$(1 - \epsilon)||v - u||^2 \le ||f(v) - f(u)||^2 \le (1 + \epsilon)||v - u||^2$.*

Essentially the spirit of the lemma is *existential*,in the sense that provides the proof about the eixstense of the mapping however it doesn't mention how it can be computed,surprisingly various computer scientists proposed methods how to achieve it algorithmically.Informally,the lemma states that always exists a mapping from R^d to R^k,where $k = O(logn)$,and we observe that this dependence is upon the number of the points and not the dimensionality per se. In practical terms this mapping can be expressed via matrix multiplication,more specifically Y=A×X,where Y$\in \Re^{k \times N}$,A$\in \Re^{k \times d}$ and X$\in \Re^{d \times N}$,and X represents the initial dataset and A the transformation or projection matrix. In the literature there have been proposed a wealth of techniques on how to populate the random matrix and is a topic of research on its own,especially in the emerging field of *Compressive Sampling/Sensing* [14].

Generally there have been proposed various methods In our case we used the following method,which was proposed from Achlioptas[10],primarily due to the very low complexity of creating the random matrix(low dependence on random bits),but more important because it leads to *sparse* matrices,hence the whole operation of matrix multiplication can be pushed further down.

$$a_{ij} = \begin{cases} +1 & \text{withprobability } \frac{1}{3} \\ 0 & \text{withprobability } \frac{2}{3} \\ -1 & \text{withprobability } \frac{1}{3} \end{cases}$$

Via this technique it can be achieved complexity of $O(dkN)$,where the initial matrix has size $d \times N$ and projecting to k dimensions,in the fortunate case where X is sparse then we can achieve $O(ckN)$,where c is the non zero entries per column.This methodology can be important in the case of streaming data and even for crafting methods for speech recognition in an online fashion.

3.3 Ensemble Classifiers-Random Forests

There is a recent trend in the machine learning community instead of focusing on crafting optimal classifiers,is better to focus on combining classifiers than may not be the optimal,the genre of these methods is also known as Ensemble Classifiers or Committee methods.One representative of this class is the method known as Random Forests[11,12], which was coined by Breiman[12].The spirit of the scheme is instead of crafting optimal decision trees which is computationally intractable(NP-complete)[15],is better to use in parallel several non-optimal decision trees and then take their majority vote(which justifies the name as commitee methods). In this section we provide some of their key characteristics,followed by the description. In particular we have the following advantages:

- Very rigorous foundations upon Probability theory
- Remarkable robustness to overfitting
- Robustness in the presence of noise and even missing data
- Relatively low time complexity

And the following disadvantages:

- Need for large memory,due to the bootsrap component of the algorithm

---

Algorithm 1. RANDOM FORESTS

1: Choose T of trees to grow
2: Pick m variables to split each node,with m≪M,where M is the dimensionality of input,m is constant while trees are developed
3: Growing of trees,when each tree we do the following:

    1. Bootstrap sample of size n by the initial set with replacement and construct a tree from this bootstrap
    2. While the tree is builded at each node pick at m variables at random and find the best split
    3. Go till maximum depth,hence no pruning is employed

4: Classify a new point via combining votes from every tree and by majority rule to decide its class label
5: With m attributes and n samples the complexity to construct a tree is $O(mnlogn)$,so if we have T trees then complexity is $O(Tmnlogn)$.Complexity is relatively low,but needs enough memory to keep all these trees in parallel and the bootstraped datasets

---

Essentially the idea behind random forests have been successful especially for datasets that exhibit high dimensionality is the following: A decision tree is a recursive partition of the high dimensional space,even if is a simple idea can be extremely powerful in practice,since trees can provide intuitive intepretation of the datasets.In parallel due to this partitioning of space it can capture non-linear dependencies among features.Recursive partition can be seen as a generalized way of *divide-conquer* type algorithm. Random forest push this idea even further and construct multiple partition of space and then by voting they can do the classification.

4   Experimental Setup

We tested our method using a subset of the phonemes of the mainstream TIMIT database. In particular we picked *17 phonemes* out of the 39 phonemes,the reason for this movement was purely computational.Essentially the spirit of our technique is to tackle phoneme recognition in its hardest case,for that purpose we merged the

Table 1. Classification of 17 classes of phonemes TIMIT using MFCC and dynamic features

Dimension	Trees	Training Time	Accuracy
48	10	50sec	53.1%
48	15	278sec	57.83%
48	20	322sec	60.89%
7	10	17.36sec	45.89%
7	40	225.5sec	50.58%
10	10	20sec	50.1%
10	20	39sec	53.31%
10	30	188sec	-crash-
15	10	61sec	52.85%
15	15	93.4sec	55.83%
15	20	146.97sec	57.61%
15	25	200.5sec	58.52%
18	10	27.09sec	54.43%
18	15	39.23sec	57.16%
18	20	58.47sec	58.88%
18	25	122.7sec	60.17%
18	30	99.3sec	**60.95%**

TRAIN/TEST folders of TIMIT into a single file and we picked equal number of every phoneme class,we took this decision in order our classifier not to be biased against misrepresented classes. Our primal goal was actually to test the ideas of the random projection method and random forests as classification algorithm and in terms of feature extraction we didn't perform and craft any fancy algorithm,more specifically we used as features the $logenergy + MFCC + \Delta + \Delta\Delta$ features with total $length = 48$.The MFCC features were computed using $frame = 256 samples, overlap = 128 samples$, $filters = 30$ We constructed a matrix of size 20995×48+1 which contains the label/class information,every class had 1235 data points. The accuracy of the classification scheme is done via k-fold cross validation,however even if is the mainstream to use as $k = 10$ folds,in our case we picked 4 folds,primarily for computational purposes which will be defined further in the following sections,but we have to report that we evaluated our scheme via cross-validation because is statistically more confident way,since we can surpass the fear of *overfitting*. The following table provides the accuracy of the multiclass-classification obtained,albeit a good way to report the accuracy of a classifier is primarily to sketch the ROC curve,but is curve only if we have a binary classification problem,in our case we would have a 17-dimensional surface(*manifold*) to 'sketch',which inherently cannot be done.

4.1 Discussion

By inspection of the aforementioned table we observe the following:

- If the target dimension is lower than the one that is dictated by the Johnson-Lindenstrauss lemma,then we see an immediate collapse in the accuracy of the classification,essentially this means practically that a significant fraction of the latent information of the initial data has been destroyed.In our case *dimensionality* = 7.

- As the dimensionality is above the critical threshold is apparent that accuracy increases rapidly,for example for dimension above 10.
- One other key aspect of our scheme is the dependence upon the number of the trees that are used to craft the classifier,when the number of the trees is increasing then also the accuracy arises.
- A general trend is that when the number of trees is increasing ,also the time needed for training rises..but we see some inconsistencies,we suspect that this happens due to temporal performance overloads on our machine(Pentium dual core 2.16Ghz)
- In some cases we report a -crash- in the table,this happens because with Weka (machine learning package) [16] we exhausted a memory uper bound of 2.5 Gbs. Unfortunately this is a problem which is based on the implementation of Weka, but also is an inherent disadvantage of Random Forests.

Overall a good aspect of the aforementioned technique is based on the simplicity,in the sense that we have very few parameters to actually tweak and experiment.

4.2 Competitive Results

In the literature there have been proposed various techniques for phoneme recognition,however there is a problem how to compare our method towards the other techniques. More specifically other researchers even apply their techniques to datasets different than TIMIT,but except this case the fundamental burden is the fact that there is no consensus how the peak the values of parameters such as window size,number of coefficients,etc. Additionally one fundamental difference is based on the fact that results are reported of accuracy based on simple partition in train/test error,however *k-fold cross validation* that is used here is statistically more rigorous and robust way to report the accuracy of a classifier. As a general trend the accuracy that have been reported is around 40%-60%.

5 Future Work

This work can be seen as the initial step towards applying techniques and ideas borrowed from high-dimensional statistics,signal-speech processing enriched with a non-linear flavor.For example it would be interesting if a modification of feature extraction method(deviating from the classic MFCC),or even a combination of features that take into account time and frequency representation of speech segments can actually improve the accuracy of the classification. Naturally, research opportunities that arise are of modifying the algorithms in streaming fashion ,or even in distributed setting,for instance Random Forests can be crafted in a parallel manner. The fact that we used random projections it opens the possibility of combining and even extending our scheme with the findings of a new emerging field of *Compressive Sampling/Sensing*[14],and in general Sparse Approximation Theory.Another possibility exists by coupling random projections with manifold learning techniques for non-linear dimensionality reduction techniques like ISOMAP[13],which seems to be more interesting mathematically but also experimentally more challenging. Other line of research can be to perform actual large scale experiments for all classes of phonemes of TIMIT and also to other datasets.

References

1. MacKay, D.: Information Theory, Inference, and Learning Algorithms (2003)
2. Vapnik, V.: The Nature of Statistical Learning Theory. Springer, Heidelberg (1995)
3. Rabiner, L.R.: A tutorial on Hidden Markov Models and selected applications in speech recognition (February 1989)
4. Mermelstein, P.: Distance measures for speech recognition, psychological and instrumental. In: Chen, C.H. (ed.) Pattern Recognition and Artificial Intelligence, pp. 374–388. Academic, New York (1976)
5. Rasta-PLP Speech Analysis Hermansky, H., Morgan, N., Bayya, A, Kohn, P.: ICSI Technical Report TR-91-069, Berkeley, California
6. Applying Large Vocabulary Hybrid HMM-MLP Methods to Telephone Recognition of Digits and Natural Numbers Krisitine W. Ma, Masters Project, UC Berkeley, ICSI Technical Report TR-95-024, Berkeley, California (Spring 1995)
7. Kokkinos, I., Maragos, P.: Nonlinear Speech Analysis Using Models for Chaotic Systems. IEEE Trans. on Speech and Audio Processing 13(6), 1098–1109 (2005)
8. Pitsikalis, V., Kokkinos, I., Maragos, P.: Nonlinear Analysis of Speech Signals: Generalized Dimensions and Lyapunov Exponents. In: Proc. European Conference on Speech Communication and Technology, EUROSPEECH (2003)
9. Fackrell, J.W.A., McLaughlin, S.: The higher-order statistics of speech signals. IEE Colloquium on Techniques for Speech Processing and their Application, 7/1–7/6 (1994)
10. Achlioptas, D.: Database-friendly random projections: Johnson- Lindenstrauss with binary coins. J. Comput. Syst. Sci. 66(4), 671–687 (2003)
11. Amit, Y., Geman, D.: Shape quantization and recognition with randomized trees. Neural Computation 9, 1545–1588 (1997)
12. Breiman, L.: Random Forests. Machine Learning 45(1), 5–32 (2001)
13. Tenenbaum, J.B., de Silva, V., Langford, J.C.: A Global Geometric Framework for Nonlinear Dimensionality Reduction. Science 290(5500), 2319–2323 (2000)
14. Compressive Sampling dsp.rice.edu/cs
15. Rivest, et al.: Constructing Optimal Binary Decision Trees Is NP-Complete. Information Processing Letters (1976)
16. Reutemann, P., Pfahringer, B., Frank, E.: Proper: A Toolbox for Learning from Relational Data with Propositional and Multi-Instance Learners. In: Webb, G.I., Yu, X. (eds.) AI 2004. LNCS (LNAI), vol. 3339, pp. 1017–1023. Springer, Heidelberg (2004)

Author Index